D0520345

If This House Could Talk...

HISTORIC HOMES, EXTRAORDINARY AMERICANS

SIMON & SCHUSTER

SIMON & SCHUSTER
Rockefeller Center
1230 Avenue of the Americas
New York, NY 10020

DESIGNED BY JOEL AVIROM, JASON SNYDER, AND MEGHAN DAY HEALEY

IMAGE RESEARCH BY ANN MONROE JACOBS

Manufactured in Italy

10 9 8 7 6 5 4 3 2 1

Library of Congress Cataloging-in-Publication Data

Brownstein, Elizabeth Smith, date
If this house could talk : historic homes, extraordinary Americans / Elizabeth Smith Brownstein.
p. cm.
Includes bibliographical references and index.
1. Dwellings–United States. 2. Historic buildings–United States. 3. Architectiture, Domestic–United States. 4. Celebrities–Homes and haunts–United States. 5. United States–Biography. 6. Dwellings–United States–Pictorial works. 7. Historic Buildings–United States–Pictorial works. 8. Architecture, Domestic–United States–Pictorial works. 9. Celebrities–Homes and haunts–United States–Pictorial works. I. Title.
E159.B8927 1999
973–dc21 99-19626
CIP

ISBN 0-684-83931-8

The author and publisher gratefully acknowledge permission to reprint material from the following works:

Pages ix and 32: Excerpts from "The Master Speed" and "America Is Hard to See" from *The Poetry of Robert Frost* edited by Edward Connery Lathem, © 1964 by Lesley Frost Ballantine, 1936, 1951, © 1962 by Robert Frost, © 1969 by Henry Holt and Company. Reprinted by permission of Henry Holt and Company.

Page 33: "On the Sale of My Farm" by Robert Frost. Copyright © 1966 by the Estate of Robert Frost. Reprinted by Permission of the Estate of Robert Frost.

Pages 86, 88, 90–93: From *His Promised Land: The Autobiography of John P. Parker* by Stuart Seely Sprague, editor. Copyright © 1996 by the John P. Parker Historical Society. Reprinted by permission of W. W. Norton & Company, Inc.

Page 240: Reprinted by permission of the publishers and Trustees of Amherst College from *The Poems of Emily Dickinson,* Thomas H. Johnson, ed., Cambridge, Mass.: The Belknap Press of Harvard University Press, Copyright © 1951, 1955, 1979, 1983 by the President and Fellows of Harvard College.

Credits and information regarding photos and illustrations may be found on page 266.

PAGE ii: *Eleanor Roosevelt was the most admired woman of her time. Her living room at Val-Kill Cottage, a converted furniture factory that was home for the last seventeen years of her remarkable life, reflects her deep interest in people and world travels.*

To the women and men throughout this land whose integrity and dedication preserve and protect the remarkable houses that tell our American story

CONTENTS

HOW *IF THIS HOUSE COULD TALK...* CAME TO BE

THE VOICE WAS EERILY FAMILIAR, BUT I WASN'T SURE WHEN I WROTE HER about my idea that I'd ever get an answer, let alone in three days. But it really was Jacqueline Kennedy Onassis on the line, and her first words were "It's thrilling... inspiring." Then her letter came. "I can't remember when I've had such a surge of excitement for the idea of a project," she wrote. Now I was sure I was on the right track.

This book must have been brewing since my childhood, for I grew up in southern New England, where houses on every street had stories to tell that made the past seem as present as today, and as stirring as current world events. I shared the terror children must have felt in their dark hiding place under the stairs of a Pilgrim's house in Plymouth as danger stalked outside. I could almost hear Longfellow reading "A Children's Hour" to his own girls in elegant Craigie House in Cambridge. When I saw the widow's walks on the roofs of captains' houses in Nantucket, I wondered how many wives, as they looked out on an empty sea, came to realize they were now widows themselves.

This book began to take shape years later during a sensational exhibition, "Treasure Houses of Britain," in Washington, D. C. I wondered what I'd list as America's treasure houses if given the chance. Happily, I was, and you are reading the result. I was sure of one thing when I began looking. I wanted houses that gave a deeper meaning to the word "treasure" than gold or glitter. My career as a researcher, writer, and producer of television public affairs programs and cultural documentaries, working with Edward R. Murrow's "boys," Eric Sevareid, historian David McCullough, and other distinguished journalists, compelled me to choose houses that could serve as metaphors, if you will, for larger events and personalities in our political and cultural history. If Abraham Lincoln was right when he reportedly said, "The strength of a nation lies in the homes of its people," which ones would tell us best? Could these houses hold any answers to today's concerns?

Finding them turned out to be a glorious adventure that took me more than once across the country, from Alaska to Florida, from Hawai'i to New Hampshire. I drove across great plains and deserts, paddled through swamps, tramped through dusty attics and ghost towns, crawled into secret compartments, and climbed staircases in everything from a shining palace to the darkest tenement. I hope some of my choices will surprise you; what I've found out may, too. Why is it so few of us have heard of a unique plantation complex designed by the greatest American architect? What is a royal palace doing in the same chapter as a condemned tenement? What is a royal palace doing among American houses anyway? What does a house that hasn't even been built yet tell you?

There are stories of great passion, courage, and brilliance in these pages, and enough cruelty, greed, and ignorance to go around. Finally, I think, the underlying message from all these houses must be what Robert Frost proposed in his poem "America Is Hard to See" about Christopher Columbus's discovery of America.

> But all he did was spread the room
> Of our enacting out the doom
> Of being in each other's way,
> And so put off the weary day
> When we would have to put our mind
> On how to crowd but still be kind . . .

Whatever your reasons for opening this book, I don't think you will soon forget these houses, the people and the times that gave them their stories. You may even come to think of your own house as a rich and worthy part of the same grand American drama.

Elizabeth Smith Brownstein

I

LIVING IN ART

Art is something most of us visit museums to admire, though our own houses contain elements that please us. In this chapter, art and house are inextricable.

For the Tlingit people of Alaska, an aesthetically pleasing, beautifully made object was inseparable from life itself. For a successful Hollywood film producer, Frank Lloyd Wright's aesthetic environments provide precious respite. David and Mary Gamble trusted two young architects to build them a bungalow that turned out to be an architectural masterpiece. For Robert Frost, the physical setting in which his farmhouse stood shaped his poetry and his conviction that there is beauty to be created everywhere, by anyone, from even the most mundane actions and surroundings.

TLINGIT CLAN HOUSE

Wrangell, Alaska

OPPOSITE: *This replica of a Tlingit clan house of the premissionary era was built in 1940 in Wrangell by the Civilian Conservation Corps.*

PREVIOUS SPREAD: *An earlier house of Chief Shakes and his clan in the harbor of Wrangell, Alaska, was photographed in the 1890s, possibly by Lt. George T. Emmons, U.S.N.*

The Tlingit people of southeast Alaska once gave their dwellings beautiful names such as

> House lowered from the sky
> Steep mountainside House
> Sun House
> Blue jay head House
> Killer whale House

One hundred and fifty years ago, thirty-one of them lined the inner harbor of a Tlingit town on Alaska's rugged, rainswept Inside Passage. Even their name for the town was vivid: it meant Human-hip lake. Americans later renamed it after the Russian Baron von Wrangel, who built a fort there in 1834. In the sheltered harbor close to shore was a small island, and on it was the Dogfish House, home of Chief Shakes, head of the Nanya.aayi clan of the Stikine Tlingits, named after the fast-running river that provided much of their livelihood in those days. George Emmons, an American naval officer, lived the last years of the nineteenth century with the Tlingit people of Sitka, a town farther up the coast. By 1882, he reported, many of those Tlingit houses ringing Wrangell harbor were crumbling away. This was far more serious than the destruction of a single-family home would be today, because Tlingit houses were communal houses, monumental structures with carefully chosen names that told of a household's origins and its history. Each house sheltered a clan, the Tlingits' principal social and political institution. The Dogfish House, for example, contained a household of perhaps fifty people: Chief Shakes's immediate family, relatives, others of lesser distinction, and slaves, all positioned on platforms in the one windowless room according to their rank in the Tlingit caste system. Slaves had to make do next to the door with the water, wood supply, fresh game, and refuse. Families marked off sleeping spaces with their mats, boxes, and screens, which could be stored underneath the platforms to form an amphitheater for ceremonies.

Seven Chief Shakes have come and gone in the last two hundred years. The first of the line won the name—and his symbol of authority, a handsome killer-whale hat now in a Seattle museum—in a fierce battle with another tribe. Chief Shakes III was a latter-day Henry V, going among his people in disguise to learn what they thought of him and all the fighting they had to do. Not long after the Russians built their fort, Chief Shakes IV moved his people up from their old settlement, Willow Town, seventeen miles south, to keep an eye on it. Chief Shakes VII, who died some fifty years ago, was the last. His Tlingit name was Kudanake, his Christian name Charley Jones. So much of the Tlingit

DOUBLE WHALE
CREST
SHAKES ISLAND
1840 - 1940
MOUNTAIN
EAGLE
TOTEM

way of life had been shattered by the time his predecessor died in 1916, a quarter of a century went by before Kudanake could be persuaded to undergo the ceremony that would formally recognize his claim to the title. No one in the present generation has taken this vital step. By 1940, few of Kudanake's people could tell the tribal stories on which the Tlingit way of life was centered. Few spoke the difficult language. Their great houses had been abandoned for the single-family homes approved by the missionaries, who were hostile to Tlingit culture. Much of Tlingit territory was now occupied by non-natives. The market economy and the end of slavery had long since upset the traditional Tlingit power structure.

Tribal elder Marge Byrd, active in the preservation of Tlingit culture, enters the House of the Bear wearing her heirloom dance robe.

But in 1940, there was a special reason for such a ceremony. On Shakes Island, a handsome replica of the chief's clan house now stood on the same spot, surrounded by eight magnificent totem poles, all part of an ambitious Depression-era restoration project in Alaska. In five years, the Civilian Conservation Corps (CCC) constructed replicas of two other communal houses, replicated fifty-four totem poles, and carved nineteen poles from memory. The 250 native craftsmen who did the work preserved some of the most remarkable art in America.

Crossing over the narrow footbridge to Shakes Island on a cold winter day is like walking into the Tlingit world of another time. Mists block out the masts of hundreds of fishing boats that on a sunny day seem to be closing in on it, just as the modern world has closed in on the Tlingit way of life. The clan house, 40 by 50 feet, with sides of thick, weathered-gray cedar planks fitted together without nails, sits low in the same pit dug for the first Chief Shakes house there a century before. Most of the totem poles on the island tower over it.

A huge brown bear in abstract form is carved on the clan house facade, and painted in black, white, and green. The brown bear is an important symbol of Chief Shakes's clan. When it was desperately climbing up a mountain to escape the Great Flood, a bear led the way, so the simplest version of the story goes. A small oval opening in the bear's belly is the only entrance, a traditional design that put unfriendly visitors in a vulnerable position. Light fixtures in the shape of ceremonial dishes were installed for practical reasons during the CCC's work and cast a low light on the painstakingly adzed walls that feel like silk.

An elaborately carved screen stretches twelve feet between two massive house posts across the wall opposite the entrance. Such a screen once served to separate Chief Shakes's quarters from the rest of his household. Behind it he slept, met privately, and kept precious possessions. The eight-foot-high screen could become a striking backdrop for the telling of clan stories by the light of the fire pit in the center of the room. Scholars are trying urgently to record Tlingit legends before the few old-time orators are gone. Until the 1960s, there was no Tlingit alphabet to preserve them in print.

ABOVE: *The 12-by-8-foot screen in a Chilkat blanket design would have separated the chief's private quarters from the rest of his clan living in the 40-by-50-foot house.*

The original screen featured the brown bear, but that treasure is now in the Denver Art Museum. The screen that replaced it during the restoration was created by the last of the Wrangell carvers, Tom Ukas, son of the master carver William Ukas. He used a Chilkat blanket design in black, yellow, and blue-green. To one scholar, such a robe was the equal of any worn by kings.

Until 1982, the new clan house sheltered another spectacular treasure that had survived two centuries of cultural upheaval, bombardment, fire, and exposure to the elements: four mammoth carved houseposts carried by canoe in the migration from Willow Town, and perhaps the oldest, most famous houseposts in the world. The function of such houseposts was to support, or appear to support, the central rafters of a clan house. The four that survived in Chief Shakes's house, with just a few traces of their original colors, measure ten feet high, almost four feet wide, and 18 inches deep. Until just a few years ago, few were aware of the name of the carver, Kadjisdu.axtc II. High-ranking clan leaders commissioned his masterpieces from around 1775 to 1800, and they were costly even then. Long before scholars identified him by name, they, too, recognized his genius and his individual style, whether it was an immense screen marking off the chief's section of a clan house, or an intricately carved dish. To Marge Byrd, a respected Tlingit elder, they are the equal of the sculpture of Michelangelo, and they overwhelm everything in the exhibit space in town that shelters them today.

Four houseposts, newly carved in 1985 and left purposely in their natural wood, have taken the place of these priceless originals. One of the carvers, Steve Brown, has since been adopted into the Kiks.adi clan. He also worked with Tlingit carvers Israel Shortridge and Gary Stevens on a twenty-foot canoe dedicated at Chief Shakes's clan house in April 1998.

BELOW: *This carved head of Gona'x'deit, a legendary sea monster, appears on one of the four 10-foot houseposts created two hundred years ago by master carver Kadjisdu.axtc II for the original clan house.*

TOP: *A Tlingit woven spruce-root crest hat shows the beauty and meaning integrated into everyday objects.*

ABOVE: *A large ladle is painted with a killer whale image, and a spoon is carved with a salmon image.*

But the House of the Bear is almost empty otherwise. "If this house could talk," says Marge Byrd, "it would like to have back the artifacts that belong here"—the original screen, the chief's staff, the drum and robe long since gone. Very simply, she explains, "To the Tlingit, art signifies who you were." Every thing the Tlingits used was made to be beautiful, or put another way by an expert admirer, "A prime requisite of all [their] art, is that it serve a useful end. In other words, a useful object such as a spoon, dish, box, paddle or weapon is made, then it is . . . decorated." The clan symbols essential to the decoration—a frog, a wolf, a killer whale, for example—would have to be dissected, distorted, rearranged to fit the size and shape of the object, which could be a sixty-foot cedar log for a totem pole, or a bear's head to be made into a ceremonial mask. The boxes, bowls, screens, blankets, tunics, hats, bows, even spoons and fishhooks—all the paraphernalia of Tlingit daily life—were so beautifully made, so aesthically pleasing, they have disappeared into museums and private collections around the world.

Even in 1879, travelers recognized the artistry of these Tlingit creations. John Muir, the exuberant American naturalist who wrote dramatically about his travels in Alaska, was sketching the totem poles at deserted Willow Town, when he heard the sound of chopping, then a thud, "as if a tree had fallen," he wrote. Someone in his party had ordered one totem pole in the deserted village cut down, and the principal figure, "a woman measuring three feet three inches across the shoulders," brought aboard their boat "with a view to taking it on East to enrich some museum or other." Chief Kadashan, the Tlingit orator accompanying Muir's party, was furious. "How would you like to have an Indian go to a graveyard and carry away a monument belonging to your family?" he asked.

Muir was impressed by what he saw among the ruins, some even then dating back a century:

> The magnitude of the ruins and the excellence of the workmanship manifest in them was astonishing . . . the first dwelling we visited was about forty feet square, with walls built of planks two feet wide and six inches thick. The ridgepole of yellow cypress was two feet in diameter, forty feet long, and as round and true as if it had been turned in a lathe. The pillars that had supported the ridgepole were still standing in some of the ruins. They were all . . . carved into life-size figures of men, women, and children, fishes, birds, and various other animals such as the beaver, wolf, or bear. With the same tools not one in a thousand of our skilled mechanics could do as good work.

The totem poles were even more exciting to Muir. Some were simple posts fifteen or twenty feet high, he reported,

Chief Shakes V lay in state in the earlier clan house in 1878, surrounded by clan treasures that include a bear mask (r) and a crest hat (l).

with the figure of some animal on top—a bear, porpoise, eagle, or raven about life-size or larger. These were the totems of the families that occupied the houses in front of which they stood. Others supported the figure of a man or woman, life-size or larger, usually in a sitting posture, said to resemble the dead whose ashes were contained in a closed cavity in the pole. The largest were thirty or forty feet high, carved from top to bottom into human and animal totem figures, one above another, with their limbs grotesquely doubled and folded. Some of the most imposing were said to commemorate some event of an historical character. But a telling display of family pride seemed to have been the prevailing motive . . . every feature showed grave force and decision; while the . . . audacity displayed in the designs, combined with manly strength in their execution, was truly wonderful. . . .

Chief Kadashan must have helped Muir begin to understand their significance; much of what we know today about the Tlingits comes from this learned chief. But even as inquisitive an observer as Muir could not comprehend the whole story just from the totem pole, any more than the chief could have understood Christmas or Easter looking at a crucifix. Subtle understandings absorbed over a lifetime were needed to understand fully what they were saying.

The eight totem poles circling Chief Shakes's clan house today remind the Stikine Tlingit people of the stories that have come down through the centuries, telling of their ancient migrations, the persistence of human failings and the tribe's need for virtuous behavior, the risks and rewards of man's intimate relationship with the elements and the animal world, the creation of the sun, moon, and stars. The poles are cultural treasures for the world, monuments in cedar comparable, some believe, to the pyramids of Egypt or the great works of art of medieval Europe. The IRA Tribal Council of the Wrangell Cooperative Association watches over Shakes Island today. The site is a precious metaphor for the Tlingits' centuries-old saga from independence and achievement, through years of decline and despair, to a new spirit of revival and appreciation.

Down the coast along the Inside Passage at the small town of Saxman, a new tribal house, the House of the Beaver, accommodates both traditional and nontraditional ceremonies. In a shed next to it, noted Tlingit artist Nathan Jackson has carved a 25-foot canoe out of a cedar log, work that required unbelievable strength and determination. It will welcome visitors from all over the world to the port, another symbol of the Tlingit's long struggle to keep their culture alive. The mystic coastal village at Totem Bight, also a CCC creation of the Depression, transports visitors back a century, another persuasive reminder that a way of life that weaves character, ceremony, history, and belongings into the art of living may in the end be a more enriching, perhaps essential, human condition than holding art in isolation, as we do in our culture.

In 1998, a twenty-foot Tlingit canoe, newly carved in Wrangell, reached Shakes Island, where it will remind visitors of Tlingit traditions. Mrs. Byrd wears her ceremonial handwoven cedar bark hat. John Martin, head of the Wrangell IRA Tribal Council, is in the rear, his wife in the middle.

FRANK LLOYD WRIGHT'S STORER HOUSE AND AULDBRASS PLANTATION

HOLLYWOOD, CALIFORNIA / YEMASSEE, SOUTH CAROLINA

IT'S NOT VERY OFTEN A MOTHER HEARS HER SMALL son say, "muv, i love you as much as i love this room!" But Pauline Schindler did, in the Storer House one day in 1931, and she loved it herself. "the room in which i sit," she wrote, "is a form so superb that i am constantly conscious of an immense obligation to mr. wright such superlative joy does it give us both. . . . "

"mr. wright" was, of course, Frank Lloyd Wright, who believed that "a love of the beautiful is the divine spark in the soul of man never to be extinguished." The Storer House was one of four houses he built in Los Angeles during a trial mid-career move to the West Coast in the mid-twenties. The move was not a success, but the house turned out to be one of Wright's masterworks.

After seven exhausting years, one of his greatest creations, the Imperial Hotel in Toyko, was finished. So, too, it seemed, was his idyllic life at Taliesin, the home he built in a beloved Wisconsin valley for his mistress, Mamah Cheney. She and her two children had been murdered there by an employee turned madman, who had then set fire to the living quarters. Wright's eldest son, Lloyd, was now a practicing architect in Los Angeles. California was booming. Surely, Wright thought, there would be a demand for a new form of architecture to suit California's climate and lifestyle. Who better than himself to create it? After all, as a fellow architect said later, "He tossed off masterpieces with the ease of a Mozart." Wright had already completed one commission in Los Ange-

OPPOSITE: *The Storer House is one of four houses Frank Lloyd Wright designed in Los Angeles using steel-reinforced concrete blocks molded in a variety of patterns.*

RIGHT: *Frank Lloyd Wright's annotated 1922 drawing of the Storer House illustrates his belief that a house should be "of the hill not on it."*

ABOVE: *This photograph of Wright at his drafting table was taken around the time the Storer House was designed.*

RIGHT: *Lloyd Wright, Wright's architect son, not only supervised construction of the house, he designed landscaping and furniture, including this table and chairs in the living room.*

les—the massive, ill-fated Hollyhock House. But by 1923, he had another kind of construction in mind. In his autobiography, Wright described it: "We would take that despised outcast of the building industry—the concrete block—out from underfoot or from the gutter—find a hitherto unsuspected soul in it—make it a thing of beauty—textured like the trees." He would mold the "outcast" on site into varied geometric patterns, eleven at the Storer House. Then, Wright said, he would become a sculptor in concrete, a weaver of these textile blocks from the warp and woof of steel rods cemented inside each block. "Then why would it not be fit for a phase of modern architecture? It might be permanent, noble, beautiful. It would be cheap."

The Storer House is the second of the four houses Wright designed in Los Angeles using that architectural Cinderella. In 1924, Dr. John Storer, the first owner, moved in, but he died a few years later. Pauline Schindler, wife of the Vienna-born architect Rudolf Schindler, a Wright protégé turned bitter competitor, was living there by 1931. Joel Silver, a Hollywood producer, bought the badly deteriorated house in 1984. It has been so carefully restored by the filmmaker together with Wright's architect grandson, Eric Lloyd Wright, that it almost seems it could have been Joel Silver, rather than Wright

himself, who said, "I believe a house is more a home by being a work of art."

Wright designed the Storer House for a misshapen lot in a treeless subdivision just beginning to feel the impact of Hollywood's expansion to the foothills overlooking Los Angeles and the Pacific Ocean. "It looked like a little Venetian palazzo, that house," Wright said. Today it nestles into a lush hillside on Hollywood Boulevard, part temple, part grotto to one observer, like the movie set of a Pompeiian villa to another. It's been a refuge from the pressures on Silver's own movie sets. "When I come home at night," says Silver, "everything's okay."

When he saw a Taliesin-designed house going up in his hometown in the mid-sixties, Silver thought, "This is remarkable, it's unique, it's spectacular. . . . When I came to California, I went to visit all the Wright buildings in L.A., and I just hoped if the day would ever come along that I made some money in the movie business, I'd be able to live in one." Well, he has, and now he is restoring a second, the little-known South Carolina plantation complex, Auldbrass.

Wright himself designed the dining room table and chairs. The gilded candelabrum is by Tiffany.

If Silver had not made the Storer House so livable, it would be tempting to say he's created a miniature museum to his idol. Over the years, he's put together a spectacular collection of original Wright pieces–chairs, tables, china, and lamps–and he's had other Wright designs custom-made from original drawings. He has complemented his Wright treasures with paintings, rugs, pottery, lamps, glassware, and flatware by superb artists celebrated during Wright's lifetime: Louis Comfort Tiffany, Charles Rennie Mackintosh, and Thomas Hart Benton. A protegée, Marion Mahoney, designed a lamp in the living room. Metal gates from a demolished Louis Sullivan office building now serve as the patio gates. Wright's son Lloyd designed the magnificent table and four barrel chairs in the living room, and the library shelves.

Lloyd also supervised construction at the Storer site, suffering all the woes that more often than not came with a Wright commission: never enough money, too much responsibility placed on inexperienced shoulders, and too many jobs going at once. At one point, Lloyd sent worried word to his absent father that the Storer House "lack[ed] joy." "Color . . . would help a lot," Wright answered. They need not have worried. "If you want good design, the basis for it is spaciousness," Wright believed, and the central core of the house is as exhilarating as the Pacific breezes that flow through it on a sunny California day. The sensation is of light, air, and vivid colors everywhere: from the blue and

yellow awnings Wright designed for the terraces on either side of the living room to the flowers in the landscaping Lloyd designed to surround the house; from the rich redwood ceiling beams to the gleaming oak floor in the living room. To Wright, wood was "the most friendly of all materials."

Pauline Schindler and her son were probably most in love with the second-floor living room, and it's easy to see why. By using what Wright called the miracle of glass, he opened the walls, "to get the outside inside the house, and the inside outside, and get the sense of freedom that goes with our Declaration of Independence," he once wrote with typical panache.

One critic described Wright as an architect-ogre, but Eric Lloyd Wright insists his grandfather had a great sense of humor. "Happy landings, always, always, always" was his response to a client who'd reported that the governor of Oklahoma had toppled out of one of Wright's notoriously unstable chairs.

The Storer House is a unique embodiment of American architectural history in another way. Eric Lloyd Wright believes it may be the only house to have had three generations of Wright architects contribute to its existence. Silver's restoration of Auldbrass adds another historically fascinating twist. In his autobiography, Wright confessed that only a handful of his many clients survived the first house and asked him to build a second. Silver and Eric Lloyd Wright came through the restoration of the first together, and have undertaken a massive project at Auldbrass, with 17,000 square feet to deal with compared to the Storer's 2,500. "It's insane compared to Storer," Silver admits. Even though he hasn't finished everything he wants to do at Auldbrass, Silver has been formally commended by the Taliesin Fellows for his "impeccable restoration of two landmark residences."

The living room of the Storer House makes it easy to see why Wright's grandson, architect Eric Lloyd Wright, calls the house he helped restore "a pavilion of light."

In 1939, when Wright was in his early seventies—with twenty more years ahead as America's greatest architect—he designed Auldbrass for a wealthy midwesterner, on the site of

Auldbrass Plantation is one of two working farms Wright designed. It was built in 1939 near Yemassee, South Carolina, when Wright was in his early seventies.

an antebellum rice plantation General Sherman destroyed on his way north from Georgia. "No other piece of architecture he did was so complete in its abstraction of nature into a structure," Eric Lloyd Wright explains. Walls lean in at an angle as gentle as do the trunks of the live oak trees that surround the complex. Copper rain spouts hang like pale green Spanish moss. Clerestory windows are outlined in an abstract pattern that might be a canoe speeding along the Combahee River with an Indian chief and his warriors, or perhaps an arrow with feather-tipped shaft.

It's difficult to imagine now that animals were living in what was left of the main building when Silver took on Auldbrass in 1986 in the nick of time. Nature and fire were finishing off the destruction the original owner's several wives had begun. "Hope dies!" wrote Wright, after he saw what they were doing to the built-in furniture and the cypress walls. Silver's collection of appalling "before" photographs makes it clear why he thinks it may take the rest of his life to complete the restoration.

Eighteen buildings make up the plantation complex today, including offices, archives, guest rooms, laundry, workshop, a six-stall stable, and tack room. Silver has several more buildings to go before all Wright's original plans for Auldbrass are realized. The largest building on the site will be almost twice the size of the Storer House. "That will be my house," says Silver, "when I've had a few more hits."

In the warm autumn sun, the Auldbrass complex stretches long and low a thousand feet along a quiet pond under towering live oaks and cypresses. In the cool of the evening, the lights in the living quarters of Auldbrass and on the ancient oak trees appear like fireflies hovering in the lowland mists. It's a breathtakingly beautiful sight, night or day.

The main house, one of eighteen buildings in the complex, is framed by a gabled portal with copper roof, and shaded by ancient live oaks.

Silver welcomes guests in the gleaming living room. Hexagonal shapes dominate here as they do throughout the complex–in the shape of the room itself, in the hassocks arranged in modules of four, even in the pattern of the Cherokee red concrete floor, waxed until it looks like leather. "The hexagon is a very fruitful unit to use," Wright once reminded his students, proposing that he was "probably the first in history to change from the square to the honeycomb." The living room is exuberant, full of flair that Wright would appreciate: orchids everywhere, a tiger rug, a jukebox, a stuffed cheetah over a desk under a Japanese lacquer screen, a tiny original model under glass of a Wright-designed reclining chair in bright blue upholstery. Silver says, "This is like Wright's own house in Wisconsin, full of personal stuff."

Plantation kitchens were traditionally set apart from the main house as a precaution against fire. Silver has made a dining room out of the long narrow pergola Wright enclosed between the living room and the kitchen. A series of hexagonal tables can be fitted together to seat up to twenty-four for dinner, or tucked into niches along one wall under the windows. The original furniture Wright designed for Auldbrass was sold at Sotheby's in 1981 for $10,000. Instead of taking up offers to buy it all back at now astro-

ABOVE: *The dining room, between the kitchen and the living room, reflects Wright's use of the hexagon in the tables and concrete floor, waxed in Cherokee red, a color that appears often in Wright's work. The sloping cypress walls reflect the tilt of the live oak tree trunks.*

RIGHT: *The spacious, informal living room of the main house continues the hexagon form, in the shape of the room as well as in the furniture.*

nomical prices, Silver chose to have everything reproduced. "The furniture was meant to be built by carpenters on site," he insists, and the rustic style of the pieces enhances the informal lodge feeling.

Silver's passion to document and fulfill Wright's plans for Auldbrass is ordered and complete, down to an original copy of the book Wright's client provided to inspire him as he created Auldbrass: *Life on a Carolina Rice Plantation in the 1850s.* A watercolor of a South Carolina cypress swamp by the author, Alice Huger Smith, hangs over Silver's desk in a frame from a Wright house in Oak Park, Illinois, where it all began a century ago.

In an uncanny echo of the feeling Pauline Schindler and her son had for the living room of the Storer House, Silver says he's happy when he sits in the Auldbrass living room. "This is the way it's meant to be." Wright put it another way: "A house should have repose and such texture as will quiet the whole and make it graciously at one with external nature." At the Storer House and Auldbrass, each man in his own way has expressed the satisfaction of "living in art."

The owner of both sites, a Hollywood movie producer, screens films for guests in the playroom at Auldbrass. The original Wright-designed furniture was sold years ago; the furniture in the complex today was made by local carpenters.

GREENE & GREENE'S GAMBLE HOUSE

PASADENA, CALIFORNIA

ABOVE: *The Gamble House is an Arts and Crafts masterpiece built in 1908 by two young California architects, Charles and Henry Greene, for an heir to the Procter & Gamble fortune.*

OPPOSITE: *The Greene brothers designed the house to connect completely with its natural environment.*

CONSTRUCTION BEGAN ON DAVID AND MARY GAMBLE'S NEW HOUSE in Pasadena, California, in early March 1908. They didn't stay around to worry about the quality of work or cost overruns, as most clients would have on such an ambitious, even radical project. Instead, they took off on a six-month trip to the Far East. Their confidence in the young architects they had chosen was justified. When the Gambles got back, not only were they able to move in a month early, they had commissioned what turned out to be one of the great houses of America.

They could afford it, for David Gamble was a rich man, son of a founder of the Procter & Gamble Company of Cincinnati. But he and his wife didn't flaunt their wealth. They were private people, conservative and pious. So they chose not to buy property near other millionaires who had discovered the pleasures of the Southern California climate, and they chose not to follow the fashion by building in one of the pretentious eclectic styles so popular in the new "Garden of Eden." It took courage, but they picked Greene & Greene, brothers whose architectural philosophy seemed to fit with who the Gambles were, and what they wanted for their retirement bungalow.

Charles Sumner Greene and Henry Mather Greene were also transplants from the East, superbly trained both as craftsmen at the experimental Manual Training High School in St. Louis, and as architects at MIT. By the time the Greenes met the Gambles, their personal vision had evolved, "to make the whole as direct and simple as possible, but always with the beautiful in mind as the final goal." This was Henry's elegant expression of the idealistic spirit that had motivated the designer-poet William Morris to found the Arts and Crafts movement in England, which reached its fullest American expression in the California of the Greenes' day, and in the Gamble house itself.

Miraculously, the house has been preserved intact for almost a century. The first narrow escape came in 1945. The Gambles' oldest son, Cecil, and his wife, Louise, were showing the house to the prospective buyers, who made the mistake of announcing they intended to paint all the wood surfaces inside the house white. Louise called off the sale immediately. She knew this would have been a disaster, because inside and out the Gamble house is a "symphony in wood," some now very rare. Every inch of wood on the walls and the furniture inside was so beauti-

fully hand-crafted and hand-rubbed to a soft finish, even Frank Lloyd Wright asked how the Greenes did it. Every tapered beam end, even the placement of each peg on the exterior structure, was as carefully worked.

Greene & Greene positioned the house to capture every breeze, every beam of light and nook of shade for maximum benefit and effect. "All art-loving people love nature first, then the rest must follow," said Charles, who shared Wright's view that "there should be in the very science and poetry of structure the inspired love of Nature." They brought flowers and trees inside the house, using wood, metal, art glass, and semiprecious stones to create natural motifs on headboards, desk legs, lanterns, tiles, and carpets.

The "dreamy" Charles, as a Morris disciple from England described him, was very much awake on site, with his architect's eye and his craftsman's understanding of the nature of wood. He was not above carving a beam himself to be sure it was shaped as he wished. When a color in his carpet designs did not turn out as he wanted in a finished piece, offending portions had to be rewoven. Every detail in the garage, the closets, the kitchen, and the attic was as carefully designed and executed as those in the principal rooms: the 38-foot-long entrance hall, the living room, and the dining room. Each piece of furniture—even the picture frames, the fireplace tools, and the switch plates—was custom-designed by the Greenes to enhance the total effect.

This wasn't pretension or conspicuous consumption. The Gambles had the resources and the taste required to sustain William Morris's injunction: "Have nothing in your houses that you do not know to be useful or believe to be beautiful." Several pieces Mrs. Gamble had collected were incorporated into the Greenes' design. A bronze crane, a motif in the Gamble family crest, hangs delicately over the settee in the entrance hall. A favorite Rookwood vase with dogwood designs sits on the master bedroom desk decorated with the same motif.

David Gamble did draw the line on the Greenes' "total environment" concept in his den. He insisted on bringing his own oak rolltop desk and Morris chair from Cincinnati, and there was nothing Mrs. Gamble or the Greenes could do to change his mind.

ABOVE: *The Burma teak entry hall staircase epitomizes the Greenes' superb use of rare woods throughout the house. Mrs. Gamble's bronze crane is integrated into the total effect.*

OPPOSITE: *Every detail in the living room fireplace inglenook, including the lighting fixture, the fireplace tools, and the carpet, was designed to fit into a unified whole.*

The dining room's built-in sideboard and the table, with an intricate extension mechanism in the pedestal, are made of mahogany. The Tiffany glass chandelier repeats the shape of the table, and the blossoming vine motif in the art glass window echoes the rosebushes outside it.

Mrs. Gamble's maiden sister, Aunt Julia, brought her brass bed to her suite, but conceded once she arrived that she really wouldn't need her Franklin stove.

The interior appears austere at first: the colors are subdued, the furniture sparse and definitely not for couch potatoes. The Gambles were indeed proper and formal. Their granddaughter, Sally, recalls her mother saying that Mrs. Gamble had wanted the furniture designed so that "you had to sit up straight and lie down flat." The Gambles disapproved of billiards, so the beautiful game room at the top of the house remained an attic. But Sally's cousin, James Gamble, who led the family campaign to give the house to the City of Pasadena in 1966, believes it is the rare and unaccustomed unity of the overall design that stuns visitors when they enter.

The living room piano is a perfect example of that unity. A grand piano would not have fitted in, but Mary was satisfied only with an upright that could produce as full a tone. So a special casing and bench were made to fit it precisely into a specific corner. "There was a reason for every detail," Henry Greene insisted, and as many times as Sally has stayed in the house, as members of the family still may do, she always discovers something new: the floral inlay of solid silver on a maple desk leg, or the way the rosebush outside the dining room echoes the abstract design of the leaded art glass window over the sideboard.

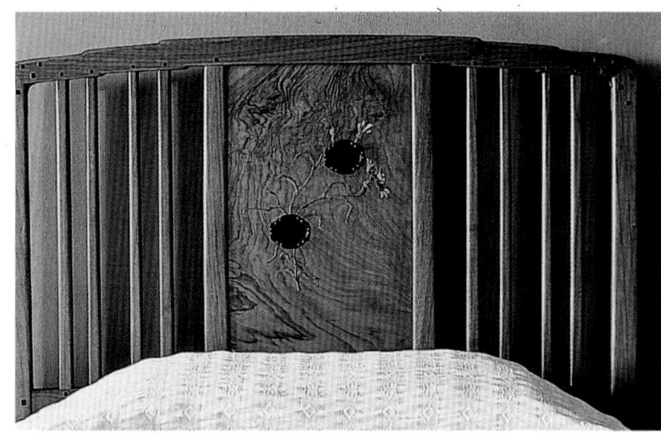

Some who know the house best remember its sensuous qualities. When Foster Gamble, a great-grandson of David and Mary, was a child, the house was a magic world to him. He recalls responding as he never had before to the feel of the wood on the entrance hall staircase. James Gamble sometimes takes the house tour incognito and relishes watching visitors quietly rub their hands over the satin-smooth paneling. A frequent houseguest confessed to creeping downstairs in bare feet at dawn to catch the sun shining through the massive teak and leaded art glass door with its California oak design, filling the house from front to back with a wonderful light.

The Gamble House cost $55,000 in 1908, twice what it took at the time to build a school in town. No amount of money could duplicate the house today. For a long time, the Gamble family didn't quite realize what a treasure they had. It was just home, possibly even a white elephant. So although the house had withstood many earthquakes intact over the years because of the way the beams are joined, it almost succumbed in 1965 to community pressures for a high-rise on the site. Today, the Gamble House is recognized internationally as a triumph of the Arts and Crafts movement in America.

People who are awed by the perfection of the Gamble House or by the originality of a Frank Lloyd Wright house may not realize how much of their own built environments has been shaped by what these three architects did. Many of Frank Lloyd Wright's clients were not rich. "Anyone can build a big house for the rich man," he said, "but design a beautiful house for the man of moderate means–ah–that will show the mettle of the architect." The Gambles were rich, but the beauty and comfort of the "Ultimate Bungalow" the Greenes created for them set a new standard for a style that spread rapidly across the country, much like the Cape Cod cottage and the ranch house.

TOP: *The design of the leaded art glass at the main entrance was inspired by the California gnarled oak.*

ABOVE: *This detail of a black walnut headboard in the master bedroom is made of semiprecious stones, ebony, and fruitwood.*

THE ROBERT FROST FARM

Derry, New Hampshire

The Frost family lived in this farmhouse from 1900 to 1909, and it "turned out right as a doctor's prescription."

On October 5, 1900, the Derry, New Hampshire, paper carried the brief bit of news that "R. Frost has moved upon the Magoon place, which he recently bought. He has a flock of nearly 300 Wyandotte fowls." A reporter looking for story ideas might understandably have concluded there wasn't much more to say there: restless young man, twenty-six, with no noticeable accomplishments, married five years, quiet wife, a baby daughter, moves to subsistence farm to try to make a living raising chickens. If he'd checked the facts, one big mistake in that simple announcement could have led him to a far more dramatic story. "R. Frost" hadn't bought the farm; he was too poor. So his grandfather came up with the money, and Frost was unhappy about that. He was also exhausted, in poor health, and worried.

It was hard for him to see a future less bleak than the past had been. His first child, a son, Elliot, nearly four years old, had just died of cholera. He blamed himself for not calling the doctor in time, and his wife, Elinor, was inconsolable. His indomitable mother was dying of cancer, and he feared for himself the prospect of the consumption that had killed his severe, heavy-drinking father at thirty-four during their years in California. He'd left college twice—Dartmouth quickly, after high school, then Harvard after two years and his marriage. He'd gone from job to job, on farms and newspapers, in factories and schools. And added to all that, the poet he had so long wanted to be was getting nowhere.

But the thirty-acre Derry farm did show promise. The tranquil setting was good for writing. The white Greek Revival clapboard farmhouse looked like new and big enough for more children. An apple orchard and vegetable garden, a neat cluster of fruit trees and bushes, and a big hayfield surrounded it. Beyond all that were the woods, the brooks and stone walls, the pastures and spring, the hills and cranberry bog, where Robert could indulge his passion for long walks and "botanizing."

The Derry farm would be home for the next nine years, quietly momentous years for Frost and for American literature. Long after he had sold the farm and sailed to England to try his fortune there, the poet wrote: "The core of all my writing was probably the five free years I had there on the farm down the road from Derry Village. The only thing we had was plenty of time and seclusion. I couldn't have figured in advance, I hadn't that kind of foresight. But it turned out right as a doctor's prescription." In the trunk he packed for England were the Derry poems he would publish successfully there as "A Boy's Will" and "North of Boston."

ABOVE: *The Frost children, Marjorie, Lesley, Irma, and Carol, found the farm as interesting a place to play as anywhere in the world.*

BELOW: *The Frost children sat on a horsehair sofa in the living room for their lessons. Their parents were experienced teachers by then.*

Plenty of time and seclusion. Today that would strike many people as a recipe for boredom, but not for the Frost family. At Derry, it had grown to six by 1905. Now Lesley, the eldest child, had two sisters, Irma and Marjorie, and one brother, Carol. (The last child, Elinor Bettina, born in 1907, would not survive.) When she was nine, Lesley wrote in the journal she had begun at five, "Our farm has interesting places to travel to, just like the world, though you do not have to journey so far as in the world." In this same small world, Robert Frost was trying to find his way as father, teacher, and poet.

The children were too far from Derry Village to go to school, so the horsehair sofa in the front parlor with the big bay window became their classroom. To Lesley, they were lucky; she wrote about "playing school." The "organized subjects," she explained later, were taught by her mother—"reading (the phonetic method), writing (then known as penmanship), geography, and spelling. My father took on botany and astronomy" . . . "before we had caught on that there was an element of *lesson* included, we were trapped."

Robert and Elinor had been co-valedictorians at Lawrence (Massachusetts) High School and were already experienced teachers. Frost was honing an education philosophy that would soon bring him growing recognition and precious income. He was demanding but compassionate, thorough yet innovative. All his life, he'd say that he wanted to "rumple [students'] brains fondly." "Think!" he wrote his daughter away at college. "Have thoughts! Make the most of your thoughts. That's all that matters." He never liked giving grades. To Elinor, marks were a curse. Frost preferred that the students help each other get A's, if only they would have ideas!

"The rule at the Frost house was you have to read by the time you're four," Lesley recalled years after, when she was helping refurnish the now historic farm with the kind of secondhand furniture the Frosts could afford in those days. By her count on April 15, 1908, the Frost bookcase held 125 books, the jewels of the house. "Reading was most important. *Learning* to read and *being read to.* Reading aloud was taught in our family. Every evening of our lives we sat before the Franklin stove in the front room, under the rather dim light of the kerosene lamps, while my mother read and my father whittled, or my father read and my mother mended." "There was poetry, poetry and more poetry, but also history, travel, drama . . . the Bible

Lesley wrote in her journal about the "rikity" stove in the kitchen; the children loved the red walls, especially at Christmastime. Frost wrote here after the family went to bed.

(many times over) . . . Darwin's *Voyage of the Beagle* . . . the Nordic myths . . . Shakespeare. Our hearts were being stretched as were our minds."

Lesley's childhood journals were published in 1969 as *New Hampshire's Child.* In her introduction, she said, "I learned . . . that flower and star, bird and fruit and running water, tree and doe and sunset are wonderful facts of life. The farm was enough of a world for learning . . . we were getting a book education by night and a do-it-yourself education by day."

The journals turn out to be a unique concordance to the poetry Robert Frost was writing at the farm. Lesley's own daughter, Lesley Lee Francis, in her book *The Frost Family's Adventure in Poetry: Sheer Morning Gladness at the Brim,* finds at least thirty Frost poems that "coincide with specific topics or incidents treated in [Mother's] daily compositions": swinging on birches, fighting bonfires, waiting for spring, watching fireflies, chasing the runaway cow, trying to sleep through a blizzard.

Frost would write at the table in the bright red kitchen long after everyone had gone to the unheated upstairs—the children to choose which bed each wanted that particular night—Elinor to the front bedroom with the rose-covered wallpaper, the only room in the house off-limits to the children. "We didn't know he was a 'poet,' or even one in the making," Lesley explained, "but we thought he was pretty good at this

Elinor and Robert Frost's bedroom was the only room the children were not allowed to enter unless they were sick.

metaphor game, at this 'imagination thing.' Those were the days when my father's poetry was ripening to the fall." Frost was discovering that poetry for him was "a way of taking life by the throat," that "inspiration doesn't live in the mud. It lies in the clean and wholesome life of the ordinary man." Life on the Derry farm was ordinary enough on the surface, made up of "chore-time–playtime–picnic time–apple-picking time–blueberrying time–hay-raking time–snow-shovelling time . . . " What Frost's Derry poems were really about were matters a child could not know: metaphors for his deep love of Elinor, his struggle from despair to hope, and his need to reconcile the cruelties of life with the existence of God.

Frost's poems were far from simple bucolic musings of a crusty, quaint New England farmer, a naive image that persists. They could not be, because they were written by a complex genius who had steeped himself for years in philosophy, history, science, literature, and the classics. They were the work of a humanist, not of a mean-spirited neurotic, the darker image of Frost that persists. Frost believed that "poetry has to heighten . . . by perpetual play of every faculty of art, imagination and figurativeness that heaven bestows. . . ." That didn't mean "loftiness," he insisted. A crow is a crow. It doesn't have to be turned into a raven to be poetry. "It's our business to give people the thing that will make them say, 'Oh, yes, I know what you mean.' It is never to tell them something they don't know but something they know and hadn't thought of saying. It must be something they recognize. . . ."

Robert Frost really wasn't a farmer, anyway. "It is not fair to farmers to make me out a very good or laborious farmer," he admitted. He hated to get up early, so he trained the cow to be milked at noon and at midnight. The farm's isolation and being at the mercy of the elements frightened him. But he did know that the land was in his bones. Years later, he wrote, "It will be a relief to get back to New England, and it's getting so I can't go without a house and barn of my own another minute longer." On the last farm he ever owned, as an elderly celebrity, he began by happily caring for a new breed of chicken recommended by a former student.

One thing is certain: Robert Frost was a devoted father who shared his learning, his imagination, and his life. One April Fool's Day, he planted colored paper flowers under an apple tree and invited Lesley for a walk past it. When she went to pick them, "they came right up out of the ground with no roots on," she reported. On Halloween, he carved jack-o-lanterns, a different face for each child. Then he hid them, each with a candle burning inside, for the children to find after the sun went down. He picked a star for each child to follow, and woke them up to wonder at the northern lights. He tested their courage by sending each child alone in the dark down the road to bring back the 10-cent piece he'd put on a box for them to find. Whoever found the perfect Christmas tree was praised. And somehow "Santer Clase" managed to leave a pile of gifts that Lesley realized only years later revealed how poor the family really was.

TOP: *Robert Frost admitted he was not a very good farmer, and trained the cow to be milked at noon and midnight.*

BELOW: *No photographs exist of Elinor during their years at the Derry farm. This picture was taken in 1911, the year the family went to England.*

Frost's involvement with his family was lifelong. His granddaughter Lesley Lee sees him as a man "who was always accessible for good talk and understanding, who presented an awesome yet human and humorous presence, and who gave unselfishly to friends and family over the years."

The Derry farmhouse itself was Elinor's "castle, province and she *was* home," Lesley remembered. There was a rocking chair in every room, because she loved to rock her babies. Her sewing room, off the dining room, doubled as a sick room because the damp and drafty house led to lots of colds and, for her husband, to pneumonia. Not long after the Derry years, Frost wrote to a friend that Elinor hated housekeeping. "Catch her getting any satisfaction out of what her housekeeping may have done to feed a poet. Rats!" Who wouldn't have felt as she did, with water for drinking and washing coming from a pump in the yard; with a stove Lesley could tell, even at seven, was "very old and worn and rickity . . . but the best stove for warmth that we have got." Elinor's health was always fragile. She was exhausted by four young children creating chaos out of order. "More than once a day the room gets into a terrible mess (with our playthings) . . . when

ABOVE: *Frost said the ax and the scythe were his favorite tools, even though they could also be used as weapons of war.*

BELOW: *Lesley Frost returned to the farm years later to help find belongings resembling those at the farm during the Derry years, including those in the pantry.*

there isn't any more room to play, then we begin to ask to go outdoors, but Mama never lets us till we have cleared it all up." But Elinor really wasn't much of a disciplinarian either. At nine, Lesley wrote a three-page drama about mischief:

> LESLEY: Mama won't like it.
>
> IRMA: O, she won't say anything; she never does; she just talks to us a little but never does any harm.

On the other hand, "Papa generally pinches us where it hurts."

And Papa could be difficult in other ways. Between the lines of Lesley's childhood stories is a man who would think up any ruse to avoid going out in the cold to bring in the cow, who was hours late getting home because he couldn't stop talking with friends in the village, and who took three times as long as he really needed to shovel paths through the snow to prove a point to Elinor. Frost could inspire love and admiration, but also fear and animosity, even in good friends.

Elinor was far more worried about money, and even sadder about life, than Frost knew while she was alive. He found out after she died in 1938, in an old letter she had written to Lesley years before. "My, my, what sorrow runs through all she wrote to you children. No wonder something of it overcasts my poetry, if read aright. I am a jester about sorrow. But she dominated my art with the power of her character and nature." To Lesley: "The power she exercised on all around her was a spiritual force, a *depth* of godliness (or goodness) that did not need God for backing–only love, human love." The inscription on her gravestone captures Elinor's role in the Frost family: "Together wing to wing, and oar to oar," from his poem "The Master Speed."

Frost lived another twenty-five years after Elinor's death, continuing to create some of the best-loved poems in the English language, translated into many more, including Persian and Arabic. Frost's poetry was shaped by the land he loved. He faced its particular New England dangers in order to create. He believed "there is at least so much good in the world that it admits of form and the making of form . . . and anybody could find it–in a basket, a letter, a garden, a room, an idea, a picture, a poem. For these we don't have to get a team together before we can play."

Just a few weeks before his own death, President Kennedy linked the poet to their beloved New England, as "one of the granite figures of our time. . . ."

> He brought an unsparing instinct for reality to bear on the platitudes and pieties of society. His sense of the human tragedy fortified him against self-deception and easy consolation. He gave his age strength with which to overcome despair. At bottom he held a deep faith in the spirit of man. He saw poetry as the means of saving power from itself . . . for art establishes the basic human truths which must serve as the touchstones of our judgment.

Frost would surely have admitted, after he had sold the Derry farm, that he hadn't the foresight then either to see what life would bring to merit the president's tribute. But an unpublished poem he wrote in 1911, when he sold the farm, revealed that he knew very well the riches he had found there:

> Well away, and be it so,
> To the stranger let them go.
> Even cheerfully I yield
> Pasture or chard, mowing-field,
> Yea and wish him all the gain
> I required of them in vain.
> Yea, and I can yield him house,
> Barn, and shed, with rat and mouse
> To dispute possession of.
> These I can unlearn to love.
> Since I cannot help it? Good!
> Only be it understood,
> It shall be no trespassing
> If I come again some spring
> In the grey disguise of years,
> Seeking ache of memory here.

Robert Frost's farm viewed from the back meadow reveals a typical New England arrangement: big house, little house, back house, barn.

II

GEORGE WASHINGTON DIDN'T SLEEP HERE

Would any of the four presidents revealed by the houses here be electable today? One suffered deep depressions, another had no charisma, and two were cantankerous and politically inept.

Yet Abraham Lincoln, a consummate politician, still holds his position as our greatest president. James Madison may have been overshadowed as president by his fellow Virginians, but he was a brilliantly effective political thinker and indispensable insider, and to Thomas Jefferson he was the greatest man in the world. John and John Quincy Adams may have had just the qualities today's cynical public might admire in a president: they pulled no punches about the state of the union, and did what they thought was best for the country even if it meant political suicide.

THE ADAMS FAMILY'S OLD HOUSE

Quincy, Massachusetts

John and Abigail Adams wanted desperately to go home. They'd had enough of diplomatic life. John had been in Europe for ten years, negotiating an end to the War of Independence, for recognition of the new nation he risked his life to bring into being, and for loans to keep it afloat. Travel for a rebel's wife was no longer as dangerous as it had been, so Abigail was finally able to join him in London in August 1784. They had spent many more years of married life apart than together, and for three or more years they'd hadn't seen their three young sons.

There never was any doubt in their minds that home meant Massachusetts. Just before they sailed in April 1788, outspoken Abigail had written to Thomas Jefferson that she was looking forward to farming and gardening, rather than "residing at the Court of St. James, where I seldom meet with characters so inoffensive as my hens and chickings, or minds so well improved as my garden." Once she and John were homeward bound, Abigail wrote in her diary, "Tis Domestick happiness and rural felicity in the Bosom of my native Land that has charms for me."

One thing at least was now certain about their future. They'd be moving into a new house. "You cannot crowd your sons into a little bed by the side of yours now," Abigail's older sister Mary Cranch had reminded her. "Mr. Adams will be employed in public business . . . that house will not be large enough for you." "That house" was the century-old saltbox cottage in Quincy, where John and Abigail had moved after their marriage in 1764. Their eldest son, the prodigy John Quincy, was born there in 1767; he and his mother had watched the Battle of Bunker Hill from a hill nearby. What seemed an ideal solution came from Abigail's uncle, Dr. Cotton Tufts. Would they be interested in buying "a very Genteel Dwelling House, and Coach House, with a Garden, planted with a great Variety of Fruit Trees, an Orchard, and about 40 Acres of Land . . . pleasantly Situated . . . about ten miles from Boston"?

For 600 pounds, a hefty sum for the Adamses in those days, the property was theirs. They had grown up, courted, married, and started their family not far from it. The seven-room, two-story, white clapboard house with a cedar-shingled gambrel roof had been built in 1731 as a country retreat by a Jamaican sugar planter. John, whose first love was farming, had always had his eye on the fertile fields that surrounded it, and there was no finer view in the world, he maintained, than its view down to the sea.

The Adamses were pleasantly surprised by the hero's welcome given the "Atlas of Independence" when they arrived at Boston on June 17, 1788. Cannons

OPPOSITE: *The garden on the west side of the Adams house provided herbs, vegetables, and fruit for two generations. It became a formal garden when the third generation had the means to turn the working farm into a gentleman's country estate.*

PREVIOUS SPREAD: *Abigail Adams died of typhoid in 1818 in the President's Bedroom in the original part of the house. John Quincy and Louisa Catherine Adams, and Charles Francis and Abigail Brooks Adams, also slept here.*

Before Abigail doubled the size of the house with an east wing, it looked to her like "a wren's house," shown here in 1798.

roared and church bells rang. Governor John Hancock invited them to stay at his official residence. John and Abigail declined his offer to escort them to their new home with a grand flourish, and it was just as well they did. When they saw the place for the first time in years, they were "most sadly disappointed," Abigail wrote her daughter Nabby, still in London with her new husband, feckless Colonel William Stephens Smith. Workmen were still swarming all over, and the furniture John had bought over the years for his diplomatic missions abroad was damaged by the hot eight-week sea voyage. Worst of all, Abigail wrote, "in height and breadth it feels like a wren's house." There were two small rooms with low ceilings–a sitting-dining room and a parlor–on the first floor to the right and left of a central entrance hall, three bedrooms on the second, and a few tiny rooms in the garret. Her idea of what was elegant and spacious had been drastically changed by the grand houses she'd known in London and Paris. "Be sure you wear no feathers," Abigail cautioned Nabby, "and let Col. Smith come without heels to his shoes or he will not be able to walk upright."

While Abigail was trying to bring order inside the house, making sure the paint colors for woodwork and walls were exactly as she specified, John was assessing the farm without much better luck. "I found my estate was falling to decay, and in so much disorder as to require my whole attention to repair it. It is not large," he wrote to an English friend. Compared to the vast Virginia estates of Jefferson, Madison, and Washington, Adams was right. But when he went on to say "It is but the farm of a patriot," he made the understatement of his long and passionate life.

Within a month, Adams's "whole attention" had been diverted by the first presidential election. George Washington, of course, would be the first president, and Adams had decided that the vice presidency of the new republic was what he wanted, and deserved. Anything else, Abigail agreed, would be "beneath himself." Adams protested often enough that politics was "an ordeal path among red hot ploughshares," but he would endure twelve more politically turbulent years before he and Abigail could "enjoy the cool Evening of Life in Tranquility undisturbed by the Cares of Politics or War," at his "sweet little farm."

Adams soon decided the vice presidency was "the most insignificant office that ever the invention of man contrived or his imagination conceived," and spent an astonishing three-quarters of each year at the farm after his first few years as vice president to be with Abigail. Even President Washington commented on his long absences. Adams regarded Washington as one of the three greatest Americans of their time, but he hated to see Washington becoming a mythical figure. Adams had analyzed wryly the "ten talents" that led to Washington's larger-than-life reputation. Adams knew he had none of them himself: a handsome face, a large fortune, the gift of silence, for example. In 1790, Adams paid the painter Edward Savage about fifty dollars for very human likenesses of the Washingtons. George had lost his teeth, and Martha Washington appears very plump and plain ("not the Tincture of Ha'ture about her," Abigail once said). At the end of his one term as Washington's successor, Adams hung their unadorned images in the formal parlor and there they have stayed, not only to commemorate the peaceful transfer of power, which was no sure thing at the time, but to reinforce his insistence that the Founding Fathers were only human. Adams knew the power of his own vanity and ambition only too well, and he had seen enough of the "gentlemen" and "simplemen" of his day to conclude "there is danger from all men." "When and where were ever found or will be found, sincerity, honesty or veracity in any sect or Party in religion, government or philosophy?" "It would seem that human Reason and human Conscience, though I believe there are such things, are not a Match for human Passions, human Imaginations and human Enthusiasms," he wrote to Jefferson.

John Adams sat for this portrait by Gilbert Stuart when he was eighty-eight. A copy by Stuart's daughter, Jane, hangs in the dining room, originally the formal parlor, and faces portraits of George and Martha Washington, which have hung in the same place since 1800.

As if to prove the point, in 1823 John Quincy commissioned the last portrait of his father; it faces the Washingtons across the room, and shows John Adams just as he was at eighty-eight, toothless, almost blind, but still ready to "Rage a little like a lion." The old patriarch had enjoyed posing for Gilbert Stuart on the red-upholstered country-style Chippendale sofa still in place in what is now called the Panelled Room, holding the cane that still rests in the stand by the front door. "He lets me do just as I please," Adams confessed, as fiercely independent in spirit then as he had been as president. He could not compromise what he thought was the right course for the country. Neither could John Quincy, when he got to the White House in 1825, as the sixth president. Father and son were the only two presidents in our history who chose to leave the capital before the

In the Book Room (also known as the study, or "the Room of Thought") above the Long Room, John Adams wrote his 109 letters to Thomas Jefferson at the inlaid oak desk, and died here on July 4, 1826.

inauguration of their successors, fully aware of their flaws as politicians, but bitter at their rejection and convinced that they had acted in the best interests of "the whole nation."

The Quincy house had doubled in size by the end of John Adams's one contentious term as president in 1801. Abigail had added a new wing to the "wren's house" without telling her husband, who was battling even his own cabinet to keep the country out of war with France, a conflict he believed could destroy the fragile new nation. To keep the roofline uniform, but to have higher ceilings, Abigail simply lowered the new wing's floor. A handsome "Long Room" now stretched 27½ by 19½ feet along the east side of the house, and above it, she created what she matter-of-factly called "the Book Room." This would become the study and cool summer bedroom for John and John Quincy, and for talented sons of the third and fourth generations: diplomat Charles Francis, historian Henry, and his brother Brooks. Out of this one room poured such a torrent of letters, histories, political philosophy, autobiographies, diaries, and poetry, that a twentieth-century admirer called it "the Room of Thought." Books were everywhere in the house, even in the hallways. Eventually, Charles Francis Adams had to build a fireproof library, made of Quincy granite, right next to the house, to contain "Every Scrap [that] shall be found and preserved for your Affliction [or] for your good," John warned John Quincy. There has never been another family archive like it. Of course, there has never been a family quite like the Adamses.

Abigail was even more pleased with what she'd done to the house after she saw Mount Vernon on a farewell visit in 1800 to her old friend Martha Washington. She reported to her youngest son, Thomas Boylston:

> My house in its present state, presents a handsome front, has larger rooms, and is better finished . . . it has more comforts and conveniences in and about it than this Huge Castle, and all I want or wish for, would be about 5000 dollars a year to spend on it, and about it.

She never got her wish. Abigail's great-grandson Henry Adams, who loved spending summers at the house when he was a child, wrote years later that it "showed plainly enough its want of wealth."

Abigail had noticed something else at Mount Vernon: it was already looking the worse for wear hardly one year after George Washington's death. James Madison's estate, and Jefferson's, also began to deteriorate after their deaths. But Abigail's house, and the few acres that remain to remind visitors of the rural setting of earlier times, would stay in the Adams family until 1946, when the property was turned over to the nation intact "with the purpose of fostering civic virtue and patriotism."

"Every room is full of history," wrote future president James A. Garfield to his wife after an overnight stay in 1869. Then an awed congressman, he added, "I doubt if any family in England or America can show such a history." Seventy-eight thousand artifacts support his point, 30,000 of them on display: family portraits, busts of illustrious friends, engravings of historic events, Abigail's bed warmer (a wedding present), the locket John gave her as he left on the first of many European missions, her daughter-in-law Louisa Catherine's prizewinning knit bedspread, the seals from the treaties John and John Quincy negotiated to end the War of Independence and the War of 1812, even the buttons from the suit a very nervous John Adams wore when he called on King George III as the first American ambassador to the Court of St. James.

Until his father's death in the Book Room, the house seemed an "abode of enchantment" to John Quincy. More often, it was "a retreat in all times of trouble," as the sardonic Henry Adams saw it. There was plenty of that. "Hot water seems our element," Charles Francis Adams concluded. From the Boston Massacre to the Mendi revolt aboard the slave ship *Amistad*, from Napoleon's defeat in Russia to the last

John Adams gave Abigail this locket in February 1778 before sailing to Europe for the first of his many long absences over the years.

echoes of the Civil War and the beginnings of the Gilded Age, an Adams was in the thick of things.

Perhaps wistfully, John Adams chose to call the Quincy house and farm

> by the Name of Peace field, in commemoration of the Peace which I assisted in making in 1783, of the thirteen years of Peace and Neutrality which I have contributed to preserve, and of the constant Peace and tranquility which I have enjoyed in this residence.

In bitter moments after his defeat in 1800 for a second term, Adams mocked himself as "the monarch of Stony Field," and the farm "Montezillo, a little hill" compared to Jefferson's Monticello, "a lofty mountain." But later on, it simply became the "Old House."

Visitors were struck by how plain and unostentatious it was, and how unpretentious and frugal was the Adams lifestyle. "Our desires are moderate, our oeconomy [sic] strict, our income, though moderate, will furnish us with all the necessaries, and many of the comforts of Life," Abigail had said in 1800, looking forward to her husband's final retirement. She did dress handsomely, visitors noticed. "Everything the best but nothing different from our wealthy and modest citizens." Guests invited to a "repast" usually described the meal as modest. Young diners were urged to eat up the Indian pudding served invariably as the first course. Whichever one managed to eat the most was told he'd be rewarded by the most generous portion of the meat. "It need not be said," recalled one young relative, "that neither the winner nor his competitors found much room for meat at the close of their contest and so the domestic economy of the arrangement was very apparent." After visiting John Adams, others were impressed that "his manner of life presented a perfect pattern for a republican chief magistrate in retirement." "A simpler manner of life and thought could hardly exist," wrote Henry Adams a century later, marveling that his grandfather John Quincy Adams was still using flint and steel to light his own fire when he arose, at 4 A.M. on the dot.

Some of John Adams's northern compatriots, Benjamin Franklin, John Hancock, and James Otis, had owned slaves to perform such tasks, but the Adamses never would, even though free labor was scarce and expensive. To John Adams, slavery was a colossal, inhumane evil, and John Quincy agreed: "It is the great and foul stain upon the North American Union." He fought doggedly against it for many years as a congressman after his presidency, despite threats to his life. So it was not surprising for Abigail to write to John Quincy, "You will find your father in his fields, attending to his haymakers and your mother busily occupied in the domestic concerns of her family," with her dog

Juno at her heels. On his first visit to the Old House in 1818, Daniel Webster was startled to find Abigail sorting the family laundry in the Long Room, while her husband read to her. "If we mean to have Heroes, Statesmen and Philosophers, we should have learned women," she'd written him long ago. A self-taught "natural genius" to her grandson Charles Francis Adams, she could carry on an informed conversation with anybody as she shelled beans for dinner. It hadn't been easy being married to John Adams. She hadn't seen him for five years when she complained bitterly to him in 1782: "I had to Act my little part alone. Tis no small satisfaction to me that my Country is like to profit so largely by my sacrifices." His years-long absences on the country's business were painful to her, destructive to the children, and to a biographer, even "unconscionable." But she remained for John "the best, dearest, worthyest, wisest friend in this world."

At home, Adams's charming, jocular side took over from the vain, irritable, pompous public figure whose "ticklish temper," "grotesque rhetoric," and "obnoxious principles" James Madison and many others couldn't stand. Thomas Jefferson could never convince Madison that John Adams was "so amiable that I pronounce you will love him if you become acquainted with him." Adams admitted he was pleased to be "descending so smoothly, the hill of life he had ascended so roughly." The retired president "had not the smallest chip of an iceberg in his composition," remembered one apprehensive young guest. "I can distinctly picture a certain iron spoon which the old gentleman once fished up from the depth of the pudding in which it had been unwittingly cooked." Adams simply continued his "good-humored easy banter." The once close friendship between Adams and Jefferson had been damaged by political differences, but they reconciled in 1812. In a stunning switch of roles, Adams, the pessimist and realist, took it on himself to try to cheer up Jefferson, once the idealist and optimist, but now disillusioned, debt-ridden, and ill at Monticello. In one of the 109 letters Adams wrote Jefferson from 1812 to 1826 at his inlaid oak desk in the Book Room, Adams urged his old friend:

> Do no wrong! Do all the good you Can! Eat your canvasback ducks! Drink your Burgundy! Sleep your siesta when necessary, and trust in God!

The mahogany paneling in the original Adams sitting-dining room, which John or Abigail had painted white, was restored by Charles Francis Adams to velvety smoothness.

LEFT: *John Quincy Adams was George Washington's prized young diplomat when John Singleton Copley painted his portrait in London in 1796. He was courting talented, beautiful Louisa Catherine Johnson.*

RIGHT: *In 1825, Louisa Catherine Adams became the first foreign-born First Lady.*

No one, of course, ever called John Quincy amiable. Neither his father, nor his son, Charles Francis, the most distinguished member of the third generation, quarreled with John Quincy's description of himself:

> I am a man of reserved, cold, austere, and forbidding manners; my political adversaries say, a gloomy misanthropist, and my personal enemies an unsocial savage.

Nevertheless, a visiting journalist, Ann Royall, whom John Quincy disliked, was astonished "by the ease and simplicity with which he received alike the lowest citizen and the distinguished stranger," sometimes supping on a piece of bread and a cup of hot water. John Quincy had never wanted to retire to the Old House as his father had, and at seventy-three he confided to the diary he'd kept since he was eleven, "Political movement [is] to me as much a necessity of life as the atmospheric air." So he ran for Congress, served eighteen years, and died during a session. "Where else could death have found him, but at the post of duty?" asked one of his adversaries.

John Singleton Copley's portrait of a debonair John Quincy hangs next to Abigail's in the Long Room. He was then twenty-eight, and in London on a diplomatic assignment, a rare moment when he was actually enjoying life, and courting Louisa Catherine Johnson. She was a gifted, delicate twenty-two-year-old beauty whose remarkable memoir, *Adventures of a Nobody,* reveals only partially the courage, the charm, and the talents she summoned up to survive a half century of marriage to John Quincy. Louisa Catherine had just arrived in the United States when her portrait in the Long Hall was painted by Edward Savage. She never quite got used to being an Adams: "Had I stepped into Noah's

Ark I do not think I could have been more utterly astonished." Louisa's hair-raising account of a five-week flight across Europe from Russia during the Napoleonic Wars, with eight-year-old Charles Francis, left no doubt that she measured up. "It was so cold," she wrote, "even the madeira wine became solid ice." Her small party kept getting lost, she was robbed, warned that her servant was a "desperate villain of the very worst character," and crossed a deserted battlefield "with an immense quantity of bones bleaching in all directions." But, unscathed, she finally reached her preoccupied husband in Paris.

John Adams came to love Louisa Catherine dearly; "wonderful woman," he called her, and she felt the same about him.

> Among all the great characters that it has been my lot to meet . . . I have never met with a mind of such varied powers, such acute discrimination . . . so intrinsically sound . . . everything in his mind was rich, racy and true.

Abigail's 1800 portrait hangs in the Long Room; she was fifty-six, and had long since proved herself "equal to every emergency in life," as Louisa Catherine put it.

Her grandson Henry Adams adored her. "She was charming, like a Romney portrait," he wrote in *The Education of Henry Adams.* He loved to watch her at breakfast in the Panelled Room, and deliver messages to her upstairs in the President's Bedroom:

> To the boy, she seemed singularly peaceful, a vision of silver gray, an exotic like her Sevres china, an object of deference to everyone.

Louisa Catherine loathed politics, but she, like Abigail, contributed mightily to her husband's lifelong devotion to public service. When she died, Congress adjourned in recognition of her impressive, gracious performance in the White House, the only foreign-born First Lady in our history. John Quincy went so far as to say, "My lot in marriage has been highly favored." Louisa admitted, "I can neither live with or without you."

The Old House was a perpetual haven for children, grandchildren, relatives, and friends, in or out of trouble. It is still the setting for the Adams family's annual picnic. Often there were sixteen or more "mouths to be filled dayly." There were weddings, even honeymoons, christenings, and dances. John and Abigail lived to celebrate their golden wedding anniversary in 1814, in the Long Room. Remarkably, so did John Quincy and Louisa Catherine in 1847, and grandson Charles Francis and Abigail Brooks in 1879. There were deaths and funerals. When Abigail died of typhoid fever upstairs in the President's Bedroom in October 1818, at seventy-four, John said, "I wish I could lay down beside her and die too." Their only daughter, Nabby, had come home in 1813 to die of cancer, having undergone what everyone hoped was a successful mastectomy at the Old House in 1811 without effective anesthesia. No wonder John Adams confessed

The Long Room in the new wing was often the scene of significant historic and family events, and contains treasures in furniture, portraits, and porcelain from four generations of a family whose political and intellectual contributions to their country are unequaled.

to Ralph Waldo Emerson in 1825, "The world does not know how much toil, anxiety and sorrow I have suffered."

But John Adams's generosity, his sense of humor, and "a heart formed for friendship" helped him cope with other family tragedies. He was eighty-nine when Ralph Waldo Emerson reported:

> He likes to have a person always reading to him or company talking in his chamber, and is better the next day we were told after having visitors in his chamber from morning til night. . . .

He listened to anything, even sermons, children's books, and the latest novels. His armchair is still in the Book Room, where he collapsed and died on July 4, 1826. He had outlived Thomas Jefferson by a few hours.

John Adams had come to appreciate the toll his long absences had taken on his boys, which may explain his delight at the prospect of having his grandchildren

> all about me. Yet they would devour all my Strawberries, Cherries, Courants, Plumbs [sic], Peaches, Pears and Apples. And what is worse, they would get into my Bedchamber [and] disarrange all the Papers on my writing Table.

The Old House had a mellowing effect on John Quincy, too, although he and Louisa Catherine had their share of sorrows, one son a suicide, another an alcoholic. Grandson Henry Adams, a rebel at six, recalled throwing a tantrum one summer day to get out of going to school. His mother was about to give in

> when the door of the President's library opened, and the old man slowly came down the long staircase. Putting on his hat, he took the boy's hand without a word, and walked with him, paralyzed by awe, up the road to town . . . until (the boy) found himself seated inside the school. . . . Not till then did the President release his hand and depart. He had shown no temper, no irritability, no personal feeling, and had made no display of force. Above all, he had held his tongue . . . neither party to this momentary disagreement can have felt any rancor, for during . . . three or four summers, the old President's relations with the boy were friendly and almost intimate.

Other Adams offspring remembered the Old House as a wonderful playground, a far cry from the time John Quincy, then six, criticized himself for spending "too much of my time at play." Henry's sister Louisa liked to kick her slippers off at one end of the Long Room, aiming them to land under her mother's handsome pier table at the other end. Her aim must have been good: Abigail's painted satin fire screen is still intact, the gilt side chairs Jacqueline Kennedy coveted for the White House are still in place, and the family portraits lining the walls of the Long Room are unscathed. That can't be said of the lives of those in the picture frames, for the Adamses demanded much of themselves. When their children were very young, John instructed Abigail:

> Train them to Virtue, habituate them to industry, activity and spirit. . . . Make them consider every vice, as shameful and unmanly; fire them with ambition to be useful. Fix their ambition upon great and solid objects, and their contempt upon little, frivolous, and useless ones.

Abigail agreed completely, and she would be pleased to know that John Adams now ranks among scholars as a "near great" president, that John Quincy earned a place in a future president's *Profiles in Courage,* and that father and son are ranked by historians just below Lincoln and Washington in character and integrity. Two centuries later, great-great-great-grandson Thomas Boylston Adams offered a contemporary interpretation of the Adamses' philosophy: "The family cared little for show. It cared a great deal for what was useful, and being itself of use to the Nation. That is what the house is saying still." The sleepy quiet look of rural Quincy has long since disappeared. Today, just inside the front door with Abigail's horseshoe still above it, John Quincy Adams's bust shifts daily with the vibrations of passing traffic, so unlike his refusal to budge politically when be believed something to be right. But Abigail's lilacs, wisteria, and roses still bloom. Her kitchen clock still ticks, and her clothesline poles are still in the backyard. John Quincy's yellowwood and black walnut trees still stand.

But what is the visitor to the Old House to make of the Adamses' dark view of human nature, or John Adamses' gloomy conclusion that "commerce, luxury, and avarice have destroyed every republican government," and he could see no reason why the United States would be any different. Now may be the very time, some scholars suggest, to pay close attention to the Adamses' message and their example. They cared passionately for their country; they served it tirelessly, at great cost to themselves and their families. They studied the past, and tried to shape the nation's future as honestly and as wisely as they knew how. They believed there was something greater than themselves. That is the larger meaning of this venerable Old House. As old-fashioned, plain, and worn as it may look, it could very well hold the answer to the troubled questions we are asking more and more about where we are going as a nation.

JAMES MADISON'S MONTPELIER

ORANGE, VIRGINIA

RIGHT: *James Madison called Montpelier, his 5,000-acre estate in the Virginia Piedmont, "a squirrel's jump from heaven."*

BELOW: *This engraving of Madison's lifelong home shows his two enlargements of the house built by his father during Madison's childhood.*

IN MAY 1794, CONGRESSMAN JAMES MADISON, FORTY-three, asked Senator Aaron Burr to introduce him to the handsome widow Dolley Payne Todd, twenty-six. Four months later, they were married. This was probably the only impetuous move James Madison ever made in his life. "My beloved, our hearts understand each other," Dolley could say after their first decade of marriage. At Montpelier, years later, one of their hordes of house guests said the old couple looked like "Adam and Eve in their Bower." Theirs was a love story every bit as powerful as John and Abigail Adams's. Montpelier, Madison's boyhood home, was where they were happiest together.

Madison's ancestors had settled in his "Obscure Corner" of America several decades before his birth in 1751. By 1760, his father had the means to build an impressive two-story, eight-room brick mansion facing the Blue Ridge Mountains to the west. Madison remembered carrying household belongings through the woods to the beautiful new site. James Madison, Sr., was more than a well-to-do tobacco planter. He was an entrepreneur, with a construction company and ironworks manned by slave labor, even doing some business as far away as Philadelphia and Kentucky. James Madison, Jr., could have settled down comfortably as a farmer on the family's 5,000-acre plantation. But his father's involvement in religious and revolutionary affairs was a powerful influence to do more. So was the enlightened education he got at college in Princeton. Madison abhorred slavery, and the prospect of managing over a hundred slaves at Montpelier troubled him. He tried to find a way to be economically independent of it, but ultimately Madison couldn't deny his inheritance as the eldest son. At Montpelier, he would even become, in Thomas Jefferson's opinion, "the best farmer in America."

But Montpelier served a deeper purpose for Madison. At twenty-two, he had concluded that some kind of involvement with politics seemed to be the answer to his search for a career. Law practice was out—"too coarse and dry," he felt—and a position in the established Anglican church in Virginia was not an option, given what would become a lifelong commitment to freedom of belief or unbelief, to the separation of church and state and "Liberty of Conscience." So he wrote from Montpelier to a college friend in 1773, soon after graduation:

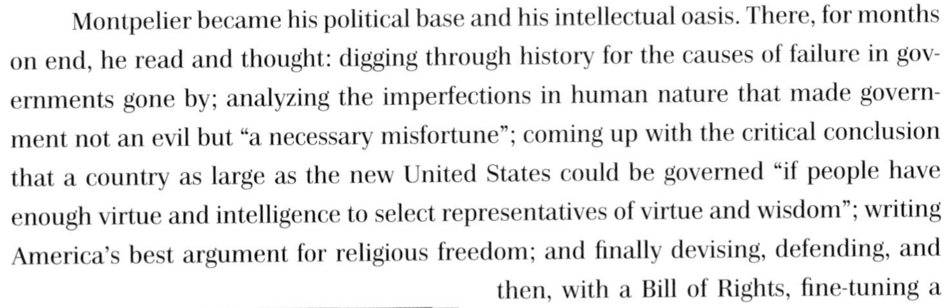

> The principles and Modes of Government are too important to be disregarded by an Inquisitive mind, and I think are well worthy of a critical examination by all students that have health and leisure.

Montpelier became his political base and his intellectual oasis. There, for months on end, he read and thought: digging through history for the causes of failure in governments gone by; analyzing the imperfections in human nature that made government not an evil but "a necessary misfortune"; coming up with the critical conclusion that a country as large as the new United States could be governed "if people have enough virtue and intelligence to select representatives of virtue and wisdom"; writing America's best argument for religious freedom; and finally devising, defending, and then, with a Bill of Rights, fine-tuning a plan to make "free Government work" despite the inevitable "mutual animosities" that would always cause friction between the people's many interests.

On the handsome front portico he designed, Madison and Dolley walked and sometimes ran for exercise in rainy weather.

A dozen times over his lifetime, Madison left Montpelier to put his political skills and plans to severe tests that would shape America's future. "If any American may be included in that small company that plays a critical role at a pivotal point in history, it was James Madison," writes historian James MacGregor Burns.

Jefferson's plantation was twenty-five miles away from Montpelier, a day's ride to the southwest. "Two brothers could not have been more intimate," said Paul Jennings, in his book *Reminiscences of James Madison.* Jennings, a slave, was Madison's valet for years. To him, Madison "was one of the best men that ever lived." To Jefferson, Madison was "the greatest man in the world." Their lifelong collaboration had begun in the spring of 1779, when Jefferson, then thirty-six, was governor of Virginia, and Madison, eight years younger, was a member of Jefferson's council at Williamsburg, the capital of the colony.

Jefferson was always "putting up and pulling down" at Monticello, but Madison's architectural approach to Montpelier was as deliberate as everything he did. Three years after he married Dolley, Madison expanded the house by thirty feet to one side, and designed an unusual four-columned, 47-foot-long portico facing the mountains. He and Dolley loved to sit there with their telescope, watching approaching visitors and the Blue Ridge in the distance. Madison's skilled slave craftsmen probably did most of the work, with a stream of suggestions from Jefferson, some heeded, some not. Then during his first term as president, Madison added one-story wings on either side, this time with the help of two carpenter-draftsmen who often did work for Jefferson. For the Madisons expected to be inundated with visitors at Montpelier after his retirement. Madison's brothers and sisters had over thirty children between them, Madison's reputation was by then international, and Dolley adored company. Company adored her, and her ice cream, even the Parmesan cheese and brown bread flavors she served. When the guest list reached ninety or more, she admitted to being nervous, but no one noticed. "They always made you glad to have come, and sorry that you must go," said one guest.

Madison was short and slight, "no heavier than a butterfly," and he dressed "wholly in black," remembered Paul Jennings, "coat, breeches, and silk stockings with buckles in [sic] his shoes and breeches. He never had but one suit at a time." Dolley, on the other hand, was famous for her wardrobe. She loved bright colors in the house . . . and on herself. Despite her severe Quaker upbringing, she "rouged" her cheeks, and later on, the curls that peeked out from her famous turbans were dyed black. And she took snuff.

Several sketches from the time show Montpelier's neoclassical facade a pale salmon color. When Madison decided to stucco the brick exterior, it's possible that iron in the Virginia Piedmont clay gave the plaster that shade, although there is new evidence he may have been advised to add "a little ochre" to the stucco. One visitor remembered Dolley's bright yellow doors. Just a hint remains of the colors inside the house: a greenish gold in the central hall, black on some baseboards, specks of green paint on a few chairs. But over the years, guests left vivid impressions. Dr. William Thornton's French-born wife found "the House . . . plain but grand . . . rendered more pleasing by displaying taste for the arts which is rarely to be found in such retired and remote situations."

Madison's love of America was the dominant theme of a huge collection of maps, prints, and portraits of ancient and contemporary notables. It covered the walls of the main rooms on the first floor of their portion of the house. Madison's mother, Nelly, lived until she was ninety-seven in the other. One of their sixty-two oil paintings mea-

sured 8 by 16 feet. Subjects ranged from "the flight into Egypt" to "a poultry yard." Busts of Founding Fathers perched on tables. Jefferson sent William John Coffee up from Monticello to sculpt terra cotta busts of Madison, his mother, and Dolley. A period cast of Dolley's was returned to Montpelier just a few years ago. It is easy to see why Madison fell in love.

The Madisons' furniture was a mix of simple Quaker, Virginia, and stylish English and French pieces. Silk damask covered walls. Blue brocade and red damask covered chairs, sofas, even the canopied bed James Monroe had bought for Madison in Paris, "from the dismantled palace of the Tuileries." Jefferson sent a marble mantelpiece. James Madison Sr.'s Chippendale-style walnut bookcase-cabinet held volumes on science, architecture, agriculture, government, religion, and history. Thousands more were piled everywhere in Madison's library. Dolley loved to play cards, so there were four mahogany card tables in the drawing room alone. Only Dolley could get away with seating political opponents at small tables for two in the dining room, although she did have a tablecloth seventeen feet long, perhaps for her big backyard barbeques. The president occasionally sat down after dinner, served from four in the afternoon to six, for a few games of whist before retiring once again to read and write. Dolley loved mirrors; one giant measured 8 by 4 feet. Madison's favorite chair was an unusual design (a campeachy chair) made of mahogany and black leather. "Everything displayed in its arrangement great order, neatness and taste," said one visitor.

Harriet Martineau, the English writer, deeply impressed by Madison's political philosophy, was invited down to Montpelier early in the spring of 1835. "Mrs. M.," Martineau wrote later in her *Retrospect of Western Travel,*

> is celebrated throughout the country for the grace and dignity with which she discharged the arduous duties which devolve upon the president's lady. . . . She is a strong-minded woman, fully capable of entering into her husband's occupations and cares; and there is little doubt he owed much to her intellectual companionship, as well as to her ability in sustaining the outward dignity of his office.

Dolley sustained Madison in many ways. Even in the darkest days of his presidency, he said he could always count on her for "a bright story and

ABOVE: *This period cast of William Coffee's terra cotta bust (1818) of the beautiful Dolley Madison was recently found and returned to Montpelier.*

LEFT: *The delicate salmon pink color of the facade may have been caused by the iron content in the Virginia Piedmont soil, or by the addition of ocher to the stucco.*

Most of the Madisons' furniture was either destroyed in the British sacking of Washington or auctioned after Madison's death by Payne Todd, Dolley's son by a previous marriage. A few rare pieces and her engagement ring have been returned.

a good laugh . . . as refreshing as a long walk." On rainy days at Montpelier, before he was crippled with arthritis, Dolley and Madison would race up and down the portico. During those rare times they were apart, she managed the plantation capably. A wistful Louisa Catherine Adams, who became a close friend, praised Dolley's "frankness and ease in her deportment that won golden opinion from all, and she possessed an influence, so decided with her little Man."

Guests were astonished to find such a cosmopolitan place in "wild and romantic" country. "His house would be esteemed a good one for many of our seats in Philadelphia, and is much larger than most of them," said an impressed Cabinet member. Yet Madison had never traveled further from home than upstate New York. Appeals to serve abroad were futile, for he always felt his health was too fragile.

There was another surprise for guests at Montpelier: a charming James Madison behind the unimposing facade one particularly nasty critic likened to that of "a country schoolmaster in mourning for one of his pupils whom he had whipped to death." Everyone conceded Madison was always "the best informed man of any point in debate," "the wisest civilian in Virginia," "the man of soundest judgment in Congress," "blend[ing] together the profound politician with the scholar." But not many knew how delightful he could be. Mrs. Frances Few's reaction was typical:

> Mr. Madison . . . is a small man quite devoid of dignity in his appearance. But a few moments in his company and you lose sight of these defects and will see nothing but what pleases you . . . his eyes are penetrating and expansive–his smile charming–his manners affable–his conversation lively and interesting.

James Madison had been married ten years when Gilbert Stuart painted this portrait in 1804, five years before he became the fourth president.

Visitors traveling from one "Sage" to the other couldn't resist comparing Madison and Jefferson. There were a few surprises here, too. Some found Madison not only "better informed" than Jefferson, but "less imperious, self absorbed," "the most profound, the most weighty." John Quincy Adams became so sympathetic to the policies of the opposition party Madison founded that his own party disowned him. He confided to his diary after a lengthy eulogy on Madison's death that he believed Madison to be "a greater and far more estimable man" than Thomas Jefferson. John Adams had told Jefferson years before, in 1817, "Notwithstanding a thousand faults and blunders, Madison's administration had acquired more glory, and established more Union than all his three predecessors." Jefferson admitted at the end of his life that he would not have accepted the presidency if Madison had not agreed to be his secretary of state.

Unlike the packed Old House of the Adamses, Montpelier today contains few traces of Madison's brilliant presence. Exhibits fill rooms that were all but empty just a few years ago. Many fine pieces the Madisons had taken to the White House were destroyed when the British burned the capital in 1814, and after Madison died in 1836, Dolley was eventually forced to sell at auction what was left. For years, Madison had shielded from her that he had spent thousands of dollars getting her disastrously irresponsible son Payne out of trouble, who then sold or squandered most of what was to be her legacy, in the days before government pensions.

Dolley moved back to Washington in 1844, and for the rest of her days she was financially strapped. Congress didn't pay nearly what she and Madison had felt his papers were worth, particularly his notes on the Constitutional Convention, the only complete and indispensable record of those proceedings. But she was beloved, and as valiant as she had been during the burning of the city. Paul Jennings was with her that day at the White House. He recalls in his *Reminiscences* that she didn't cut down Washington's portrait herself. "She had no time for it. It would have required a ladder to get it down. All she carried off was the silver in her reticule, as the British were thought to be but a few blocks off." The White House doorkeeper and gardener removed the picture for her. One spot at Montpelier remains much as Madison knew it: the small round garden temple he built around 1811 near the mansion. At first, it's hard to reconcile this delicate, serene place with Madison's hardheaded political savvy, and a lifetime spent at the center of the new Republic's political ferment. Then again, he was as tough and cool a politician as they come, and perhaps there is some symbolism in his construction of the temple over his ice house, which rose over his father's buried ironworks.

Montpelier is alive today with Madison's spirit, the center of efforts to assure Madison the place he deserves in the American pantheon. John F. Kennedy may have started it with his remark that Madison is "the most underrated of our Founding Fathers." He was the last of them as well. Madison, the Father of the Constitution, outlived the Atlas of Independence and the writer of the Declaration of Independence by ten years. Harriet Martineau left a glowing report of her conversations with the eighty-three-year-old Madison at Montpelier over several scintillating days in 1835:

> He was in a chair, with a pillow behind him, when I first saw him; his little person wrapped in a black silk gown; a warm gray and white cap upon his head, which his lady took care should always sit becomingly; and gray worsted gloves, his hands having been rheumatic. His voice was clear and strong, and his manner of speaking particularly lively, often playful. . . . He seemed not to have lost any teeth, and the form of the face was therefore preserved, without any striking marks of age. It was an uncommonly pleasant countenance. . . .

They talked for hours: "on the subject of slavery, more than any other," about population growth (Malthus had just died), about ancient Roman agriculture, international trade, literary copyright, education ("no distinction in this respect should be made between men and women"), freedom of religion, philosophy, international relations, Darwin's poetry, national debt, and taxes. Martineau was captivated and hated to leave. She knew her English readers, "living as they did under institutions framed by the few for the subordination of the many," were keenly interested in Madison and Jefferson, "men inspired by the true religion of statesmanship, faith in men, and in the principles

The garden temple, close to the mansion, was built by Madison over his ice house, which was built over his father's iron-works. It is the purest symbol of the presence of the brilliant, charming, and underappreciated James Madison.

on which they combine in an agreement to do as they would be done by." Madison, she concluded, "reposed cheerfully, gayly to the last on his faith in the people's power of wise self-government."

And she reminded her readers that Jefferson, in his last letter to Madison, had written, "To myself you have been a pillar of support through life. Take care of me when I am dead, and be assured that I shall leave with you my last affections." Since Madison's Constitution has been the pillar of support for us for over two hundred years, that seems the least we can do for him.

ABRAHAM LINCOLN'S HOUSE

Springfield, Illinois

This may be the most poignant spot in all America—the only home Abraham Lincoln ever owned. The solid "Quaker brown" wooden dwelling seems too small to contain the sadness—and greatness—that draw people to it from all over the world. Most know Lincoln would never return here once he became president. He was assassinated less than a week after the end of the Civil War, in April 1865. But few know the full extent of the Lincoln family's tragic story. One of the four Lincoln boys, Edward Baker, died in the house in 1850, after a fifty-two-day struggle with pulmonary tuberculosis. Two of the three sons who survived childhood here, the beloved Willie and loyal Tad, would not live to adulthood. Lincoln's wife, the still controversial Mary Todd Lincoln, would finally be destroyed by what was to come.

ABOVE: *This photograph of Lincoln was taken on October 1, 1858, during his debates with Stephen A. Douglas. He was forty-nine.*

OPPOSITE: *Lincoln's Quaker brown house in Springfield, Illinois, was the only one he ever owned, and by 1860 it was the most impressive dwelling in the neighborhood.*

When Lincoln left home as president-elect on a cold, wet February morning in 1861, seven Southern states had already seceded from the Union. Ten days later, Jefferson Davis would be sworn in as president of the Confederate States of America, soon to number eleven. No wonder Lincoln told the crowd seeing him off at the railroad station: "I leave . . . with a task before me greater than that which rested upon Washington."

In calmer days, Lincoln had said to a friend, "It isn't the best thing for a man . . . to build a house so much better than his neighbors'," but by 1860, he had. Stunning discoveries by the Lincoln Legal Papers Project have now revealed how he was able to afford it. Lincoln may have started out as a poor, self-taught country lawyer, but by 1860 he was one of the most prominent, successful attorneys in the West, even admitted to practice before the Illinois and United States Supreme Courts. "He could split hairs as well as rails," says one historian. His legal practice had provided most of the $1,500 to buy a one-and-a-half-story Greek Revival cottage on the edge of town in 1844, two years after he and Mary Todd were married. A backbreaking load of cases, 5,000 in all—from petty theft to murder to corporate liability—tried in crude log cabin courtrooms and ornate federal court chambers, had enabled him and Mary to keep on enlarging the house. Fees, for one case as high as $5,000 but usually around $50, laid the financial base for their political and material ambitions.

When politicians and the press converged on the house in the summer of 1860 to take a closer look at the new Republican Party's presidential nominee, the house was ready and Mary Todd Lincoln was confident her husband was ready, too. As a girl, she'd said she would marry a president one day. She had chosen Lincoln over other suitors, including his debating and presidential rival, "the Little Giant" Stephen A. Douglas, because she was sure of Lincoln's greatness the moment she met him. In 1858, Lincoln

By 1860, the house had been enlarged to accomodate a growing family and the Lincolns' political ambitions. Lincoln, still without beard, stands behind the picket fence with Willie, and on the fence is his youngest son, Tad.

joked to a young journalist, "Mary insists . . . that I am going to be Senator and President of the United States, too. Just think of such a sucker as me as President."

He was one of the shrewdest and most effective politicians of his day and had helped organize the Republican Party in Illinois. In those days, as Lincoln said, "the man who is of neither party is not, cannot be, of any consequence." But the most he'd ever managed was a two-man law office, and a ragged band of Black Hawk War volunteers. Compared to the Adamses and Madison, Lincoln's qualifications for the presidency were unimpressive. Mary's family certainly felt that way, but she knew better. Later she would boast, "He rose grandly with the circumstances of the case, and men soon learned he was above them all." One of the Lincolns' neighbors believed that "without Mary Todd Lincoln for his wife, Abraham Lincoln would never have been President." Mary's warmest supporters and her harshest critics agree on that if nothing else.

Two stories in the *New York Herald* in August 1860 carried descriptions of the Lincolns' house:

> It is like the residence of an American gentleman in easy circumstances, and is furnished in like manner. There is no aristocracy about it; but it is a comfortable, cosy home, in which it would seem that a man could enjoy

> life surrounded by his family. . . . The internal appointments clearly show the impress of Mrs. Lincoln's hand, who is really an amiable and accomplished Lady.

Much maligned in her day and still controversial, Mary Todd Lincoln was in her early forties in this photograph. Some believe Lincoln would not have become president without her.

Mary's "impress" begins at the sidewalk. Instead of the wooden planks in front of the other homes in the neighborhood, the Lincolns' sidewalk was paved with brick, laid in a herringbone pattern. One reporter noticed that "the house is built plumb to the sidewalk," so there was hardly any front yard for Mary to embellish. Today one lonely elm sapling grows out front, just as one did in that famous photograph of Lincoln, still beardless, standing with two of his "codgers" at the corner of the house, in 1860. The back yard was hopeless—the barn, the woodshed, the privy, the fruit trees, and garden took up most of it. Lincoln's horse and the cow grazed in an open field beyond the woodshed and privy.

There is no way the interior could have been anything but Mary's, either. Lincoln was as indifferent to his surroundings as he was to his dress. He'd had to be, to survive his harsh, dangerous frontier childhood. He had nothing but the clothes on his back, and two saddlebags, when he moved to Springfield in April 1837. He was already twenty-eight years old, with an odd assortment of jobs behind him—surveyor, postmaster, riverboatman—and just one year of formal education—defective, he called it. Mary, on the other hand, was the daughter of a wealthy Kentucky aristocrat, Robert Todd, of Lexington. She had been raised in comfort and style—with slaves, politically influential family friends, including Henry Clay, and a superior education.

Mary knew she and Lincoln were "of opposite natures." But their house at 8th and Jackson is a monument to the many things that kept the two devoted to each other until the end: their ambition and fascination with politics, their love of children and friends, and their passion for poetry, theater, and books. Mary's sister, Elizabeth Edwards, never could understand what Mary saw in Lincoln, but she could tell that he "was charmed with Mary's wit and fascinated with her quick sagacity—her will—her nature—and culture." Mary was tiny, affectionate, pleasingly plump, vulnerable, "the very creature of excitement" to another lawyer in town. Even Lincoln's law partner, William Herndon, whom Mary would not allow in the house, conceded she was "witty and fearless." Her brother-in-law, Ninian Edwards, said, as a young belle, "she could make a bishop forget his prayers." Mary agreed that Lincoln was "not pretty," but, she said, "the people are perhaps not aware that his heart is as large as his arms are long." "I would rather marry a good man—a man of mind—with a hope and bright prospects ahead for position—fame & power than to marry all the houses—gold . . . in the world."

Mary Lincoln's bedroom adjoined her husband's in the fashion of the time and is a perfect illustration of the Victorian decorating philosophy of "harmony through contrast."

Behind the plain walnut front door with the simple nameplate "A. Lincoln" is the front hall, with a stairway up to the five bedrooms created with their remodelings over the years. Lincoln irritated Mary when he opened the front door in his shirt-sleeves or bed slippers. Neighbors remember him lying on the hall floor with his feet against the newel post, dangling one of the boys or a pet above him. Lincoln and Mary were indulgent parents for their day, even ours. After they'd gone to Washington, Lincoln's barber reported, "Mr. Tilton (their renter) has no children to ruin things." When Lincoln served youngest son Tad the first helping, he told a dinner guest, "Children have first place here, you know." Mary Lincoln agreed when he said, "It is my pleasure that my children are free—happy and unrestrained by parental tyranny. Love is the chain whereby to lock a child to its parents." Mary described her childhood as "desolate." When she was six, her mother died, and her stepmother produced nine rivals for her father's affection in fifteen years. Lincoln had lost his brilliant mother, Nancy, when he was nine. He loved his stepmother, Sarah, and what she had done to bring order and gentility into his young life. But his early years had been desolate in their own way. "It is a great piece of folly to attempt to make anything out of my early life. It can all be condensed into a single sentence you will find in Gray's *Elegy*; 'The short and simple annals of the poor. That's my life, and that's all you or anyone else can make of it.'

Whether you turn right off the front hall, to the family's sitting room, or left into the front and rear parlors, the wild mix of colors and patterns in wallpaper, drapes, and carpets astonishes twentieth-century eyes. The Lincolns' upstairs bedrooms, separate but connecting, in the fashion of the day, are just as exuberant. Harmony through contrast was the Victorian decorating scheme, and Mary Todd Lincoln followed it as far as she could afford to, and sometimes beyond, for she always had trouble managing money rationally. The most fashionable furniture, walnut and mahogany Rococo Revival, was reserved for the formal front parlor, the only room off-limits to their unruly boys. (Mary also had to lock the cupboard in her tiny kitchen to keep them out of her desserts.) Lincoln's study in the back parlor, and the rest of the house, contained a mix of Empire-, Gothic-, and Cottage-style pieces.

They sold or gave away most of their furniture when they moved to the White House. A number of carefully documented treasures have come back over the years: Lincoln's shaving mirror (he was clean-shaven in the Springfield years), their small dining room table, a wood-burning kitchen stove, and the battered mahogany lap desk he carried in his saddle-bag when he rode the circuit. Lincoln sometimes conducted business in his bedroom, sitting on the four-poster bed that probably was lost in the Chicago fire of 1871. Perhaps it was this practice, Mary's migraines, and his nightmares that led to adjoining bedrooms. But in her pitiful last days, Mary slept only on one side of her bed, so as not to disturb "the President's place" beside her.

Lincoln's bedroom was on a front corner of the house. He sometimes conducted business here.

The Lincolns entertained hundreds of people here over the years. Five hundred guests were invited to one enormous party in 1857. Mary reported to her half-sister, Emilie Todd Helm, that "owing to an *unlucky* rain, 300 only favored us by their presence . . . [at this] very large and I really believe very handsome & agreeable entertainment." A "strawberry company" for seventy in June 1859 was modest in comparison. Mary even sent out fifty handwritten invitations to Willie's tenth birthday party! Lincoln sometimes wrote them himself.

The most significant gathering in the house was the Lincolns' reception of the delegation that came to notify him formally of his presidential nomination. One member of the committee wrote:

> About six o'clock in the evening the Republican Convention Committee called, and after the usual salutation, . . . Mr. Lincoln standing on the threshold of the back parlor and leaning somewhat on an armchair—the Committee formed before him in the front parlor . . . Mr. Lincoln looked

In the formal parlor, on May 19, 1860, Lincoln was officially informed of his nomination for the presidency.

much moved and rather sad, evidently feeling the heavy responsibility thrown upon him. He replied briefly, but very pointedly. . . . All appeared to have a foreboding of the eventfulness of the moment, and all felt that in this contest there was more than the mere possession of power and office at stake, nay, the vital principle of our national existence.

We know from drawings in *Frank Leslie's Illustrated Newspaper* of March 9, 1861, that Leonard Volk's plaster bust of Lincoln, from a life mask, still without the beard, was in the front parlor on a whatnot. "There is the animal himself," Lincoln would say. Seeing Lincoln for the first time could be startling. He was six-foot-four, awkward, lanky ("A skeleton in a suit," someone said). Mary, five-foot-two, would never pose with him. They were "the long and short of it," he teased. Partisan critics, even in the North, were merciless.

"Obscene clown . . . long-armed ape . . . an awful, woeful ass" were just a few epithets. But more often than not, a negative first impression changed dramatically after a Lincoln speech or encounter. One skeptic heard Lincoln's speech at Cooper Union, and left thinking he was "the greatest man since St. Paul." Lincoln's White House secretary, John Hay, who saw him day after day during the brutal, desperate first years of the war, wrote:

> The boss is in fine whack. I have rarely seen him more serene & busy. He is managing this war, the draft, foreign relations and planning a reconstruction of the Union . . . all at once. There is no man in the country so wise, so gentle and so firm. I believe the hand of God placed him where he is.

Mary Todd Lincoln has never been treated as kindly. She was a complex, difficult woman, who could be cutting, hot-tempered, stingy, possessive, volatile ("either in the garret or the cellar," said one acquaintance). She panicked easily, especially when Lincoln was away riding the 8th Judicial Circuit for weeks at a time. Running a cramped household with help as transient as it was in those days, and handling their rambunctious boys ("the dear rascals," Lincoln called them), wasn't easy for a high-strung, proud woman who'd been brought up in the lap of luxury. When she married Lincoln, she'd had to learn how to cook, make the family's clothes, and keep house herself—with no indoor water supply, a privy in the back yard, candles and whale oil for light. "Her hand is not soft," noticed a visitor to the White House.

Lincoln couldn't have been the easiest man to live with, even though he sometimes did the dishes and shopping. His marriage may have been "a matter of profound wonder" to him, but Mary realized her husband "was *not* a demonstrative man, when he felt most deeply he expressed the least." He could be as stubborn as she was. "He was a terribly firm man when he set his foot down," Mary said. "No man nor woman could rule him after he had made up his mind." He was infuriatingly absentminded, not noticing for perhaps a block that one of his boys had fallen off the wagon he was pulling. When he let the fire go out because he was absorbed in a book, after three reminders Mary hit him on the nose with a stick of firewood. He could never be counted on to be home in time for dinner. "Bring on the cinders," he said once when he arrived two hours late. In the hundreds of gallant letters she wrote in the years of exile in Europe after the assassination, Mary remembered his "deep amiable nature," his "boyish mirth." He was "always . . . lover—husband—father & *all all* to me—Truly my all."

Lincoln remained true to her. Once, in the White House, Lincoln turned to a guest and said, "My wife is as handsome as when she was a girl, and I, a poor nobody then, fell in love with her, and what is more, I have never fallen out." They were holding hands

Mary learned to cook in this tiny kitchen. She was raised in a well-to-do Kentucky family with slaves, and she chose to have a dining room rather than a larger kitchen.

when he was shot. That very day, Lincoln had said to Mary, "We must *both* be more cheerful in the future. Between the war and the loss of our darling Willie, we have both been very miserable."

The house at 8th and Jackson was where, for seventeen years, they had their best days. Mary could not bear to return to it. "Occupying the same rooms, breathing the same air, where so many happy years had passed–the contrast without my husband would simply deprive me of my reason," she told a friend in 1866. Finally, Mary had to go back to her sister's house in Springfield to die, after years of physically painful, mentally anguished wanderings through Europe and the United States. Her loyal neighbors, Noyes Miner and his wife, found her one day in New York in a small "dimly lighted and plainly furnished room in New York . . . almost blind . . . partially paralyzed . . . weak and alone."

It is a wonder she was alive and rational at all. Her mother was not the only family she lost when she was young. A brother had died when she was four. Her father died of cholera in July 1849, her favorite grandmother in January 1850, son Eddie a month later, Willie in 1861, three brothers and a brother-in-law in the war, and Tad in 1871. Her husband had been murdered beside her, after years of worry over gruesome threats to his life. For marrying Lincoln, Mary was hated by Southerners as a traitor to her heritage, and attacked by Northerners as a Confederate collaborator. She saw herself as a scapegoat for both sides. The press and Washington society treated her cruelly as First Lady. She made many foolish and costly mistakes in judgment, but her social successes and her good deeds went unreported. Mary was embittered by the generous treatment of other public figures whose sacrifices she felt did not compare to hers. She believed she was betrayed by her seamstress and close friend, Elizabeth Keckley, a former slave, who published a book about her years with the Lincolns in the White House. The ultimate betrayal came from her only surviving son, Robert. He regretted for the rest of his life that he had arranged to have her tried for insanity in 1875. We know now, from "The Insanity File" Robert kept hidden in a closet in his Vermont mansion, Hildene, that was the only way under Illinois law he could legally get control of her finances. This he wanted urgently to do, in order to protect her from her own irrational obsession with money, and the ill will of her political enemies. Within a year, a court hearing found her "restored to reason," but the damage to their relationship, and what was left of her reputation, was irreparable.

Lincoln's ranking as our greatest president holds. His compassion, his humanity, his eloquence, and his martyrdom have brought him immortality. In the grandest tribute of all, Leo Tolstoy believed that Lincoln was "what Beethoven was in music, Dante in poetry, Raphael in painting, and Christ in the philosophy of life. He aspired to be divine—and he was."

The best Mary could do was the testimony of some of her neighbors. James Gourley told Herndon, "I don't think that Mrs. Lincoln was as bad a woman as she is represented; she was a good friend of mine . . . a good woman." Noyes Miner, in his *Vindication of Mrs. Abraham Lincoln,* testified after her death that she was "a devoted wife, a loving mother, a kind neighbor, and a sincere and devoted friend. Blessings on her memory."

New medical analysis of Mary Lincoln's physical and mental condition, as diagnosed by four eminent physicians just six months before she died, proposes that she was an untreated diabetic, which caused a chronic spinal cord disease, locomotor ataxia, with numerous symptoms very similar to those she herself described vividly, but which were interpreted by many at the time as indications of her insanity. Perhaps the most compassionate epitaph came at her funeral in Springfield. The Reverend J. A. Reed compared the Lincolns to two "lofty pine trees with branches and roots intertwined. Though only one was struck by lightning, they had virtually both been killed at the same time. With the one that lingered, it was slow death from the same cause."

III

THE TRUTH ABOUT TARA

The "magnolia myth" is still alive, but the houses here expose the harsh realities behind it.

Rosedown is the quintessential antebellum plantation house, and would have made a perfect Tara. Behind its gleaming facade was a life that brought little joy to the plantation mistress who ran it, and none to the slaves who worked it. Two houses in Ripley, Ohio, were perilous first stops on the Underground Railroad, that invisible road thousands of courageous slaves traveled toward freedom. Booker T. Washington was born a slave, but the house students built for him on his Tuskegee campus became a symbol for freedmen coping with the brutal aftermath of the Civil War.

ROSEDOWN PLANTATION

St. Francisville, Louisiana

OPPOSITE: *The "Big House" at Rosedown was built by slaves in 1835 for Martha and Daniel Turnbull. It is the centerpiece of one of the most comprehensive plantation complexes remaining in Louisiana.*

PREVIOUS SPREAD: *An avid amateur horticulturalist, Martha Turnbull planted the seeds for this quarter-mile oak alley leading to the plantation house through her extraordinary twenty-eight-acre formal garden.*

COTTON BROUGHT A FINE PRICE IN 1835, THE YEAR Daniel Turnbull's slaves finished building a grand "Dwelling House" on his plantation in remote, beautiful West Feliciana Parish, Louisiana. He and Martha Hilliard Barrow had been married six years almost to the day he noted in his journal: "commenced hauling timber cypress for the new house." On their honeymoon in 1828, the Turnbulls had seen a romantic play, *Rosedown*, and when they chose that name for their new mansion and the 3,455-acre cotton plantation surrounding it, they were following the custom along the River Road, but places like Belle View, Golden Grove, and White Castle were equally unrevealing of what plantation life was really like. The new house was finished on May 1, 1835, the cost: $13,109.20. Both the tally and the simple penciled original floor plan are still on Daniel's library desk. The small ivory disk still embedded in the newel post of the spiral staircase in the entrance hall assured visitors the house was paid for.

Not that there was much question about the Turnbulls' financial status. That same year, 1835, Daniel sold over 170,000 pounds of cotton for as much as 18½ cents a pound—and that was just his crop at Rosedown; his three other plantations totaled 5,550 more acres. An envious young brother-in-law, Bennett Barrow, confided to his diary, "Saw Daniel Turnbull in town today . . . fat, his pockets full of money." He and Martha hadn't had to start from scratch as a married couple; both brought substantial inheritances in land, slaves, and cash.

Fortunes could be made fast along the Mississippi in those days. Entrepreneurs came from the Upper South, the North, even Europe, to try their luck. By the 1850s, there were said to be more millionaires on the River Road between Natchez and New Orleans than in all the rest of the country. But financial ruin could come just as quickly. An ornate plantation house downriver, built on sugar money, was called San Francisco, a corruption, we're told, of "sans fruscins"—without a penny. Several other sugar planters chose "Hardtimes" for their operations. They used to say it took a rich cotton planter to start as a poor sugar planter. Cotton really was king, and Rosedown was one small slice of a vast plantation system in the Deep South that produced 79 percent of the nation's cotton—over 5 million bales just before the Civil War began. The March 1858 issue of *Harper's Magazine* described the situation eloquently: "Cotton is on the levees, cotton is on the steamers, and cotton is in the mouths and bosoms of all the people." It had become the country's leading export, worth more than all other exports combined. Textile factories in England were eating it up, and New England was industrializing on the strength of it.

Cotton was king throughout the South, picked by gangs of slaves, supervised here by a fellow slave, the driver. Slaves were sometimes imported from specific regions in Africa for their expertise in growing a particular crop. The task system was used in rice cultivation.

Thousands of slaves were moving into the region each year from the worn-out lands of the Upper South to work the fields. As many as a million men, women, and children were shipped down between 1790 and 1860, a forced migration as momentous as the first, from Africa to America, that began in the seventeenth century.

Daniel owned 450 slaves, which made him one of the very few Southern planters owning over 200. Fewer than 1 percent had even a hundred by 1860. "The influence of large owners must have been enormous," suggests historian John Hope Franklin in his classic *From Slavery to Freedom,* "since they have been successful in impressing posterity with the erroneous conception that plantations on which there were large numbers of slaves were typical . . . the bulk of the slave owners were small farmers . . . more than 200,000 owners in 1860 had five slaves or less . . . 338,000 owners, or 88% of all owners in 1860, held less than 20 slaves." Three-quarters of the white people of the South had none, mainly because they couldn't afford any.

Daniel and Martha might have been surprised to learn that their splendid mansion would reinforce the "magnolia myth," that persistent image of an idyllic life of luxury, leisure, docile and beautiful belles, dashing gentlemen, and happy, devoted slaves. Myths die hard, and what Frederick Law Olmsted criticized at the time as "the deluge of spoony fancy pictures" of plantation life is one of the most indestructible. If David O. Selznick had wanted an authentic antebellum plantation setting for *Gone With the Wind,* Rosedown was a made-to-order Tara. He needn't have settled for that plywood facade with papier-mâché columns on MGM's back lot. Scarlett could have flirted effectively on the cool galleries running the full 60-foot length of the first and second floors on the front of the house. The magnificent alley of live oaks dripping with Spanish moss was perfect for dramatic entrances by her beaux. The "full gretio doric collumns [sic]" Daniel's slaves carved would have completed the picture of the plantation ideal.

In reality, very few planters lived in houses remotely like Rosedown. Fanny Kemble, the brilliant and beautiful British actress married for a time to the wealthy owner of a rice plantation in Georgia, described their own house

> as rather more devoid of the conveniences and adornments of modern existence than anything I ever took up my abode in before. It consists of three small rooms, and three still smaller, which would be more appropriately designated as closets, a wooden recess by way of pantry, and a

kitchen detached from the dwelling–a mere wooden outhouse, with no floor but the bare earth . . . our sitting room a bare, wooden-walled sort of shanty. . . .

But Daniel had instructed his contractor that "the workmanship and stile [sic] of the new house were not to be surpassed in the state." He and Martha saw to it themselves that the furnishings were also unsurpassed. They collected superb Regency and Empire pieces for the dining room. The finest American cabinetmakers of the day, Crawford Riddle and Anthony Quervelle of Philadelphia, and John Belter of New York, created elegant furniture for their bedrooms, the parlor, and the card room. Jonas Chickering built their piano and music stand. Thomas Sully painted portraits of Daniel and Martha, and of two of their three children, William and Sarah. Daniel's impressive library contained rare folios as well as books inscribed to his "darling daughter." Spectacular French-made scenic wallpaper covered the walls of the entrance hall. A statue of Venus by Canova, a rare family portrait by artist John James Audubon, their neighbor for a time, and a Nubian clock, were of museum quality.

Martha Turnbull's portrait by Thomas Sully hangs in the dining room. Most of Rosedown's original furniture is still in place.

The Turnbulls also installed the latest modern conveniences. Daniel and Martha even owned a "refrigerator," made of rosewood and mahogany outside, lead inside, and filled with chips from huge blocks of ice floated down the Mississippi from the North. They installed a shower in the Greek Revival Henry Clay wing, which they'd added to the original structure after the election of 1844, to accommodate a mammoth Victorian Gothic bedroom set friends of Henry Clay had planned for him to take to the White House. Daniel wanted to show his admiration for Clay, even if it meant having to add a matching wing on the opposite side of the house to maintain the symmetry of the facade. The couple imported hundreds of rare and exotic trees and plants for the remarkable 28-acre formal garden Martha created out of the wilderness, inspired by the gardens she had seen in Europe.

LEFT: *Daniel Turnbull's library was impressive. The original floor plan of the house, and the bill, are still on his desk.*

OPPOSITE: *This mammoth Victorian Gothic bed was made for presidential hopeful Henry Clay to take to the White House. Daniel had to build a new wing on the house to accommodate it, and a matching wing on the other side.*

But the reality of plantation life at Rosedown was quite different from the beautiful images that can still be conjured up there, even though Martha's description of her housewarming party only seems to reinforce the "magnolia myth."

> We had 30 people at our first party & we had 6 chickens for Chicken Salad—2 turkeys, 2 Ducks, 1 Ham, 1 Tongue, Roast Mutton. 2 roast Chickens, 1 Pig—Henrietta took 12 dozen eggs and made a great deal of cake. . . . It appears useless to make so much cake—2 Neuga Ornaments costs 74 dollars—Musicians 60 dollars indeed to induce everything it cost 224$.

Such lavish hospitality was only a small part of what any plantation mistress had to contend with every day, even one as wealthy as Martha. She hinted as much in the garden journal she kept for sixty years. "I've had so much company I have kept a poor account of my garden," she wrote in 1853. A few years later: "I had to stint from having company so long." Well-bred as she was, this is as far as Martha cared to go, it seems, to describe what her life was really like. On the traumatic events in her life, at least before the Civil War, she said little. Life was supposed to appear graceful, effortless.

Margaret Mitchell gave a partly accurate glimpse of the realities in *Gone With the Wind* when she described how dilapidated Tara had become since the death of Scarlett's mother. Gerald O'Hara's "sharp blue eyes noticed how efficiently his neighbors' houses were run and with what ease the smooth-haired wives in rustling skirts managed the servants. He had no knowledge of the dawn-till-midnight activities of these women, chained to supervision of cooking, nursing, sewing and laundry. He only saw the outward results, and those results impressed him."

Everyone and everything was vulnerable to the whims of nature. "The climate is a wretched one and destructive of human life," Thomas Jefferson was told in 1804 by the Louisiana governor, whose family had been decimated by a yellow fever epidemic. The Turnbulls' seven-year-old son, James Daniel, died of yellow fever within hours in 1843. On October 7, 1860, an overseer reported, "We were all getting sick and I believe that an [epidemic] of some kind would have been inevitable had we been compelled to use such impure filthy water much longer." "Very dry and parching, many things dying," Martha wrote in her garden journal on June 18, 1838; "Too wet to plough" (July 8, 1853); "Going on 4 weeks since the Sun shone it is freezing all the time. It is nearly exhausting to everyone to be so pinched up with cold" (January 3, 1856).

"I find," wrote a contemporary of Martha, "by daily experience I am of a hardier mold than I had the most distant idea." If Martha Turnbull had any doubts about how hardy she was, even as a fledgling plantation mistress not long out of her teens, the view from any window at Rosedown would have reminded her of responsibilities few in her position could afford to think about tomorrow. Vast quantities of food had to be grown, prepared, and distributed to plantation workers from the Cook House right behind the "Big House" in the work yard. The accidents and illnesses of hundreds of people had to be attended to with the primitive instruments and remedies of the day. The slave hospital is long gone, but the plantation doctor's office was just a few yards away from her front door. *Gift for Mourners* was the disheartening title of one of the books on his desk! A church and a barn for slave "frollicks" have disappeared too, and nothing remains of the slave quarters, which family records indicate were "laid off on the plan of a small city," not too far from the doctor's office. Historian Anne Firor Scott adds to the workload:

> No matter how large or wealthy the establishment . . . fine ladies thought nothing of supervising hog butchering . . . or of drying fruits and vegetables for the winter. They made their own yeast, lard, and soap, set their own hens, and were expected to be able to make with equal skill a rough dress for a slave or a ball gown for themselves. It was customary for the mistress to rise at five or six, and to be in the kitchen when the cook arrived to overlook all the arrangements for the day. . . .

The Turnbulls loved music. Their piano and music stand were made for Rosedown by Jonas Chickering. A rare family portrait by John James Audubon hangs above the piano.

No owner despaired of slavery more than James Madison, but in his conversations with journalist Martineau at Montpelier about the time Martha was moving into her new house, he lamented, "the saddest slavery of all was that of conscientious Southern women." "For women as for men," historian Scott writes, "slaves were a troublesome property." To Deborah Gray White, the plantation was "akin to a psychological battleground where slaves vied with the whites in a never-ending clash of wits."

Sojourner Truth, born a slave in 1792, surely would have scorned Madison's view, and she described the cruel predicament of slave women in a stunning appearance at a woman's rights convention in Akron in 1851. An unknown slave, Sarah Grudger, echoed Truth's words years later.

> I nebbah knowed whut it wah t'rest. I jes wok all de time fom mawnin' till late at night. I had t'do ebbathin' dy wha t'do on de outside. Wok in de field, chop wood, hoe cawn, till sometime I feels lak mah back sholy break. I don ebbathin' cept split rails . . . Ole Marse strop us good effin we did anythin' he didn' lak.

The plantation doctor's small office was not far from the Big House.

Historian Sally McMillen concludes that "most [Southern women] whatever their status or color, led difficult and exhausting lives" and "Life was a stream of hardships sprinkled with a few moments of joy."

As the war began, Martha Turnbull was managing Rosedown on her own. Daniel, a veteran of the War of 1812, died in the fall of 1861. Her eldest son, William, had drowned in 1856, in his mid-twenties. A bust of William in the front hall, and the black mourning stripes on the portico of each wing, were constant reminders. Only her daughter Sarah was left. They stuck it out at Rosedown together for the rest of their lives, with Sarah's husband, James Pirrie Bowman, the boy next door, and their ten children. Sarah's marriage was not a happy one. "I will not make sacrifices for a man who does not love me," was one of her milder comments.

Thousands of documents in the Turnbull family archives deal with the complexities of running their large commercial operation, far fewer with personal matters. A clue to the scale of the enterprise Martha had to take on lies in the claims she filed after the war, for restitution of property taken from the Turnbull plantations by the Union Army and Navy in just one month, June 1863: "300 hogsheads of sugar, then worth $60,000; Six hundred barrels of mollasses, then worth $18,000; Two hundred head of mules then worth $30,000 . . . " and there was much more. Her estimate of the total value of the losses that one month came to $183,000. In 1902, Sarah was still trying to collect: "I only know that the four plantations were absolutely stripped of every possible thing," she testified.

The inscription on an old sundial in Martha's beloved garden reads, "I measure only the happy hours," but after 1863 there were very few. "Since the Federals landed in May," Martha noted in her journal in January 1864, "neither field or garden has been worked. The garden is a wilderness it looks melancholly [sic]. Nothing to eat." By then, well over a hundred Turnbull slaves had deserted to the Union Army. Freedman Joseph J. Harris, now a sergeant in the Union Army in Florida, wrote his general:

> to ples to Cross the Mississippia River at Bayou Sar La with your Command & jest on the hill one mile from the little town you will finde A plantation Called Mrs. Marther H. Turnbuill & take away my Farther & mother & my brothers wife with all their Children. . . . I will amejeately Send after them. I wishes the Childern all in School. it is beter for them then to be their Surveing a mistes. . . .

Those former slaves who stayed at Rosedown had changed, too. Even loyal Augustus, whose name appears in Martha's journal for twenty-nine years, balked at her orders. "Augustus said he would not cut wood to put in my wood house when I told Ben to tell him to do it. Simon would not weave. For nine days, Lucinda refused to come and wait on me. . . ." One thing did work in Martha's favor; the Big House was several miles from the Mississippi, so at least it escaped shelling by Union gunboats. There is evidence that Federal troops did encamp on the plantation as early as 1863. Even if Rosedown had been on the river, Sarah's prewar national reputation as a spirited beauty might have prevented its destruction, if the many stories are true that Union officers, confronted by lovely prewar acquaintances on the steps of their threatened mansions, ordered their troops to spare them.

The disintegration of a way of life was still underway a decade later. There was little cash to hire former slaves, who now had to be paid. And many freedmen did not want their wives to work at all, especially not in the fields. Martha wrote in her journal August 23, 1872, "Cleaned up my [work] yard entirely by my own hands," and on April 17–18, 1874, "Not one Negro in the field & only made one contract today—terrible trouble. It looks like a famine. I do not know what to do." Her last journal entry, over twenty years later, on September 1, 1895, was just as pitiful. "My pension came. I had not one dime to pay Emma $2 this month, August, nor any debt whatever. . . ."

At Rosedown today, there's barely a hint of this chaos, but a photograph of Martha taken on the front porch a few months before she died at eighty-seven in 1896, shows the toll the hardships took. Old films record what was left of the decaying house in 1955, when Martha's last granddaughter, Miss Nina, died. The antebellum grandeur was long gone. Four of Sarah's ten children, unmarried daughters, unable to find suitable husbands after the loss of so many men during the war, had hung on at Rosedown after she died in 1914, using the fireplaces for heat, cooking on an 1880s wood stove, with only three lightbulbs in the entire house, and one hand-cranked telephone. They were destitute, and survived by selling off some land, cuttings from what was left of Martha's garden, a few chickens, and the contribution from an occasional curious tourist. The granddaughter of one of Martha Turnbull's slaves had helped them pull through. Amazingly enough, they left

Rosedown debt-free, basically intact, and still in the family.

But in 1956, Rosedown finally had to be put on the market, and the ideal buyer came along by chance. Catherine Fondren Underwood, of Houston, Texas, was as passionate an amateur horticulturist as Martha. She could see through the decay in the house and the ruin in the famous garden, to their original splendor. She also had the financial resources to handle a meticulous eight-year restoration of the house, the grounds, and some outbuildings. The result: one of the most significant, rarest, and beautiful plantation complexes in the South. Over 85 percent of what the Turnbulls themselves used in the house is still on the site today, and many of her plants and trees survive, including the magnificent alley of oaks leading up to the house.

At Rosedown, it's deceptively easy to slip back into that mythic antebellum world, because the plantation is still insulated from the harshness of the present landscape along the Great River Road by 1,800 acres of woodland, pasture, and Martha's re-created formal gardens. But we resist, in deference to the labors of all who worked there, and with the help of the current owner, who believes that the reality is more enriching than the romance.

ABOVE: *Martha died several months after this picture was taken in 1896. She was eighty-seven, and had managed to hang on to Rosedown through the Civil War, Reconstruction, and economic depression.*

RIGHT: *Scenic wallpaper like this was purchased by the Turnbulls in France for the graceful entrance hall at Rosedown, now restored to its original splendor.*

THE UNDERGROUND RAILROAD HOUSES OF JOHN PARKER AND REVEREND JOHN RANKIN

RIPLEY, OHIO

JOHN PARKER'S TWO-STORY BRICK HOUSE STOOD right on the northern bank of the Ohio River, in the town of Ripley, Ohio. On the same side of the river, looking down on the town, the river, and the Kentucky shore from the top of a high hill, was John Rankin's house—also brick and two stories. The two houses were no more than half a mile apart, but getting from Parker's to Rankin's was about as perilous a journey as anyone could take in the free states of the Union before the Civil War if you were a runaway slave or, like Parker and Rankin, a conductor on the Underground Railroad. "How near we may be to safety and yet lose all," John Rankin wrote years later, when it was no longer dangerous to reveal one's role in the operation of that clandestine, invisible road. Ohio state law, and Federal law as far back as 1793, made it illegal to help escaping slaves in any way, and the Fugitive Slave Act of 1850 made matters even worse. At that point, said John Parker, "Everyone engaged in the work destroyed all existing evidence of his connection with it."

Rankin wrote his autobiography in 1873 when he was eighty. Parker dictated his life story to a journalist friend, Frank Gregg, when he was an old man, probably in the late 1880s. Their memories of those violent, tense times were still vivid. "I am now living under my own roof," said Parker, "which still stands just as it did in the old strange days. . . . It too has heard the gentle tapping of fugitives. It also has heard the cursing at the door of the angry masters. It too has played its part in concealing men and women seeking a haven of safety. . . . We have seen adventurous nights together, which I am glad to say, will never come again." "My house was in full view of Kentucky," Reverend Rankin recalled. "The slaves by some means discovered that I was an abolitionist, and consequently, when any of them ran away, they came to my house. . . . I have had under my roof as many as twelve fugitive slaves at a time, all of whom made good their way to Victoria's dominions. My house has been the door of freedom to many human beings, and while there was a hazard of life and property, there was much happiness in giving safety to the trembling fugitives." His Underground Railroad work went on for forty-three years, from 1822 to 1865. Parker's first experience with runaways came in 1845, and his involvement continued through the end of the war. He even recruited troops in Kentucky for the Union Army's 27th Ohio Volunteer Infantry (Colored) Regiment.

By the time Eastman Johnson painted A Ride for Liberty: The Fugitive Slaves *around 1862-1863, over 100,000 slaves had fled the South, as many as 40,000 through Ohio alone.*

NEAR RIGHT: *The John Parker House was declared a National Historic Landmark in 1997, as was the Rankin House. Parker, born a slave, had purchased his freedom after several unsuccessful attempts to run away himself. Diane Tweedle (in the yellow slacks), a direct descendant of John Parker, talks with Charles Nuckolls, Reverend James Settles, and Miriam Zachman, leaders in the campaign to make the house into a museum.*

FAR RIGHT: *John Parker was an inventor and foundry owner, as well as a fearless conductor on the Underground Railroad. This is one of his three patents.*

OPPOSITE: *Reverend John Rankin's house overlooked Ripley, Ohio ("a hell-hole of abolition"), the Ohio River, and Kentucky, a slave state. Lights in the bedroom windows signaled slaves to make a run for it across the river.*

Both men were solid citizens to those in Ripley who opposed slavery as vehemently as they did, but "fierce passions swept this little town and divided its people into bitter factions," Parker remembered. Spies, slave catchers, decoys, and traitors were everywhere, in town and the surrounding countryside. To them, Ripley was "a hell hole of abolition." Parker always walked in the middle of the street to avoid being jumped by his enemies from the narrow alleys that riddled the town. He was always armed, and often in disguise when he was operating.

Parker was more than familiar with what fleeing slaves were going through, because he had tried several times to run away himself. The child of a Virginia aristocrat and a slave woman, he'd been sold first by his father in Virginia when he was eight, later marched with other slaves in chains to Alabama, where he was sold again. Parker managed to buy his freedom in 1845 for $1,800 (plus interest), then headed north. He was eighteen. By 1859, Parker was the owner-operator of a successful iron foundry next to his house on Front Street. The single-piece, three-step cast-iron stair still leading to his front door was made in his foundry. Patents for three successful mechanical inventions came later.

John Rankin was a Presbyterian minister with a reputation as an abolitionist extending far beyond Ripley. He'd moved to Ohio from Kentucky, a slave state, so he could preach against slavery and work for emancipation with less risk to himself and his family, or so he thought. A collection of thirteen letters he had written in 1824 to his

slave-owning brother in Virginia about slavery's evils was published and became required reading for antislavery forces across the country. One admirer, William Lloyd Garrison, called Rankin the "Martin Luther of the cause."

Parker said he'd never heard of Ripley until his first, reluctant participation as a conductor on the Underground Railroad landed him there after ferrying two young slaves across the Ohio River from Maysville, Kentucky, the town ten miles upriver. The two girls waiting on the riverbank were "unusually fat," Parker noticed, and he soon learned why: they had made off wearing their mistress's hoop skirts, each with three or four of her dresses on top. And they were loaded down with bundles as big as they were. "Any other time I would have laughed, but serious and dangerous work to me was at hand," Parker recounted years later. The trail they left, of discarded dresses, satin slippers, even a frying pan they planned to use to cook their bacon on the road, plus the "swath they cut through the cornfield as big as a bull elephant could make," almost doomed their escape, said Parker, but from then on, he was hooked by the "excitement of the chase," and the "adventures that required all my skill and resourcefulness."

Parker did not want to risk hiding fugitives in his own house; it was far too vulnerable, "easy of access to anyone who cared to row across the river and walk up to the top of the bank. It was also remote from my friends, which made it all the more accessible to those who ventured to attack me." So he kept the front door "locked and chained so no one could enter." That way, he figured, if anyone knocked at the door, Miranda, his wife, "a good watchdog herself," could open their bedroom window, "while I stood beside her to size up the situation." Parker knew there was a price on his head in Kentucky, because he had seen the sign himself: REWARD $1000 FOR JOHN PARKER DEAD OR ALIVE. For Parker did not limit his conductor work to the Ohio side of the river. He went back into Kentucky again and again to get runaways started, sometimes reluctantly, on their escapes to Canada.

Once when Parker did let runaways into his house it was a near disaster. Two fugitives had appeared at his door toward dawn, "too late to get them away before daylight," Parker realized. So he had to hide them in his attic, "hoping there was no one following the fugitives to my house." His hope turned to despair when an armed crowd rushed him at his front door and proceeded to search every room and closet. Parker was sure they would eventually find the runaways. "I could see the confiscation of all my land and seized property, and the wreckage of my whole life's work," he said, but his

TOP: *The Parker House stood right on the Ohio River bank, and by 1910 had fallen victim to floods and fire. Parker helped hundreds of runaways over twenty years.*

ABOVE: *This "Freedom Stairway" led up to the Rankin House on Liberty Hill. A sturdier structure has replaced it, each step named for a contributor to its construction.*

delaying tactics had given the two men time to climb up to the roof and hang on for dear life until the slave hunters left.

At the Rankin house, the door was always left unlocked. Lights in second-story windows signaled to runaways on the Kentucky side of the river that if they could somehow manage to get across, by swimming, rowing, or walking when water was low in summer or iced over in winter, they could expect help from the Rankins and friends both black and white to get out of Ripley as fast as the Underground Railroad could carry them. Parker described Ripley as "the real terminus" of the Underground Railroad, which got its name there when a runaway disappeared so quickly after crossing the river his frustrated owner, in hot pursuit, was told the slave "must have gone off on an underground road."

John Parker preferred to work on his own, but he knew the Rankins well. "The real fortress and home to the fugitives was the house of Reverend Rankin perched on [a] high hill behind the town," he recalled. "If shadows on the wall could but return, you could count the sire, with six sons, seven resolute men, holding their border castle against all comers. At times attacked on all sides by masters seeking their slaves, they beat back their assailant, and held its threshold unsullied . . . in this eagle's nest, Reverend John Rankin and his sons held forth, during many stormy years. . . ." Parker came to the Rankins' rescue at least once, armed with his double-barreled shotgun.

Two thousand escaping men, women, and children hid in the Rankin House on their way to freedom in Canada on the Underground Railroad.

In those days, Rankin's house had a bedroom wing, a barn with a secret cellar, and sheds to serve as hiding places. The plain, sturdy, six-room house today stands alone on Liberty Hill as a symbol of the Rankins' simple belief that "no man could love his neighbor as himself and yet hold him in a position in which he would not himself be held." To Parker, "slavery's curse was not pain of the body, but pain of the soul." "The real injury was the making of a human being an animal without hope. It was the taking away from a human being the initiative of thinking, of doing his own ways."

Reverend Rankin also had a price on his head in Kentucky—$2,500. His house was under constant surveillance, and Rankin knew that "men lay around my house at night to murder me." A letter to a local newspaper was probably of small consolation to his family. "I would remind my friends from across the River that if they murder Mr. Rankin, destroy his property, and even burn the town of Ripley, they will never save one slave by it."

Twelve fugitives at a time could be hidden in the Rankin House, in sheds and in a secret cellar in the barn.

From 1845 to 1865, Parker helped over 440 runaways on their way to Canada. He had stopped counting at 315, when he realized it would be wise to burn his records in the foundry furnace. As many as 2,000 men, women, and children passed safely through the Rankin house over the years. One of his sons said modestly, "All that my father did in the aid of fugitives was to furnish food and shelter. His sons . . . did the conveying away." Mrs. Rankin and her four daughters shared fully in the risks; so did Parker's wife and their children.

As family men, both Parker and Rankin were especially moved by poignant stories of mothers or fathers risking everything to go back into slave territory to bring out the rest of their families. Parker told of a single man who gave himself up to bounty hunters and a doomed future, so that the husband of another escapee might have the last seat in Parker's skiff to be with his wife on their flight to Canada. Both knew the agony of the mother who, like William Styron's Sophie, had to choose between leaving one lagging child behind, or stop and be captured with her husband and other child.

Harriet Beecher Stowe made literary and political history from one fugitive's story she heard at the Rankin house. For many years it was called "the Eliza house" because the model for her dramatic version in *Uncle Tom's Cabin* first found shelter there with her child, after escaping across the melting ice on the Ohio. "When her husband finally found her," Reverend Rankin wrote, "she became so frantic that she could scarcely believe it was he. She shouted aloud and manifested the most violent passion. Spring came and they got safely to Canada."

Reverend Rankin was called "the Martin Luther of the cause" by abolitionist William Lloyd Garrison. Rankin's family shared fully in the dangers of conducting on the Underground Railroad, as did John Parker's.

Until fugitives made it to Ripley, and other such stations along the Railroad, they had to rely pretty much on their own luck, intelligence, and courage. William Still, son of a runaway slave, and for fourteen years a conductor in Philadelphia, described what it took, long after the Underground Railroad stopped running:

> Some were guided by the North Star alone, penniless, braving the perils of land and sea, eluding the keen scent of the bloodhound as well as the more dangerous pursuit of the savage slave-hunter; some from secluded dens and caves of the earth, where for months and years they had been hidden away, from mountains and swamps, where indescribable suffering from hunger and other privations had patiently been endured. Occasionally, fugitives came in boxes and chests, and not infrequently some were secreted in steamers and vessels and in some instances journeying hundreds of miles in skiffs. Men disguised in female attire, and women dressed in the garb of men have under very trying circumstances triumphed in thus making their way to freedom. . . .

Parker never forgot the taste of the soup a slave cook left for him, without saying a word, when he was trying to escape. "Life took on a new aspect as I left behind me that friendly house and the kindly old cook and the fragrant bean pot."

As in other epic struggles, there is poetry. "My soul is vexed, my troubles are inexpressible," wrote Isaac Forman, who'd been separated from his wife by sale before fleeing. "I often feel as if I were willing to die. I must see my wife in short, if not, I will die. What would I not give no tongue can utter. Just to gaze on her sweet lips one moment I would be willing to die the next." Jermain Loguen, also an Underground Railroad passenger, and later a minister and Underground Railroad conductor, wrote: "No day dawns for the slave. Nor is it looked for. It is all night—night forever." The lights shining from the Rankin house were one of the first signs that day might dawn.

THE OAKS OF BOOKER T. WASHINGTON

Tuskegee, Alabama

The Oaks was the fifteen-room brick residence of ex-slave Booker T. Washington, founder of Tuskegee Institute, which opened on July 4, 1881.

On the day emancipation came in April 1865, Booker T. Washington was only nine, but years later he described the epochal event as if it were yesterday, and as only Booker T. Washington would. He and the other slaves gathered at the "Big House," Washington remembered:

> . . . we were told that we were all free, and could go when and where we pleased. . . . For some minutes there was great rejoicing and thanksgiving, and wild scenes of ecstasy. But there was no feeling of bitterness. In fact, there was pity among the slaves for our former owners. . . . Within a few hours . . . I noticed . . . there was a change in their feelings. The great responsibility of being free, of having charge of themselves, of having to think and plan for themselves and their children, seemed to take possession of them . . . Was it any wonder that a deep feeling of gloom seemed to pervade the slave quarters? . . .

Soon after, Booker T. Washington, his mother, brother, and sister left their former owners' small Virginia farm for West Virginia to join Booker's stepfather, who had belonged to the owner of another plantation and saw his family rarely. He'd run away to the Union Army during the war, as thousands of slaves had, and after freedom came, he'd found a cabin for the family, and jobs for himself, Booker, and his brother, at the salt furnaces in the small town of Malden.

It would be hard to imagine a cabin worse than the one Booker's family had lived in as slaves. That was one small room made of split oak logs with openings in the wall for windows, a dirt floor, piles of rags for beds, little food, fewer clothes. But the freed family's house in Malden *was* worse. It was in town, crowded together with other cabins, surrounded by decaying garbage, stinking outhouses, and a "motley mixture of the poorest, most ignorant and degraded" neighbors, he discovered. Washington never liked town life after that. He would always be a Southern country boy, suspicious of city people, their airs, and esoteric knowledge. Later he would say, "No race can prosper till it learns that there is as much dignity in tilling a field as in writing a poem."

Washington's life as a slave was nothing out of the ordinary. His world was limited to the plantation, his white father did not acknowledge his existence, he had no shoes until he was eight, and no schooling (he taught himself to read). But Booker T. Washington was "no ordinary darkey" to General Samuel Chapman Armstrong, the founder of Hampton Institute, who was soon to become Washington's lifelong inspiration and

905

Construction of the Queen Anne–style house by Tuskegee students was supervised by an architect on the Tuskegee faculty, one of the first black graduates of MIT.

supporter. And by the time Washington's autobiography, *Up from Slavery,* was published in 1901, when he was in his mid-forties, he had left the squalor of Malden far behind. He had worked desperately hard to get an education, and at twenty-three, he had become the founder and principal of Tuskegee Normal and Industrial Institute, which went on to become one of the nation's most highly acclaimed African-American educational institutions. He was now world-famous, his luck-and-pluck uphill struggle in perfect harmony with the Horatio Alger spirit of the Gilded Age. He had just finished building The Oaks, a handsome Queen Anne–style two-and-a-half-story red brick house with parlor, library, dining room, kitchen, family and guest bedrooms, den, breakfast room, five bathrooms, a spacious veranda on three acres of garden, orchard, and pasture.

One of Washington's Northern benefactors—a Philadelphia Quaker businessman—wondered why it had to be quite so big. "When at Tuskegee lately," he wrote, "I noticed that a *very large house* was being built. . . . If it is thy house . . . this seems hard to reconcile with thy position and the needs of the School. . . ." Washington asked his most trusted adviser, William H. Baldwin, Jr., to defend it, which wasn't difficult. The Oaks was far more than an official residence and haven for the indefatigable Washington, his redoubtable wife, Margaret Murray Washington, and his three children, Portia, Booker Junior, and Ernest Davidson. It was a model, a symbol. "I have found," he wrote in *Up from Slavery,* "that it is the visible, the tangible, that goes a long way in softening prejudices. The actual sight of a first-class house that a Negro has built is ten times more potent than pages of discussion about a house that he ought to build, or perhaps could build."

This was the essence of Booker T. Washington's pragmatic philosophy. He had learned from experience, he wrote, "that there is something in human nature which always makes an individual recognize and reward merit, no matter under what colour of skin merit is found." Washington's new home, The Oaks, was only one "visible, tangible" piece of evidence he presented at Tuskegee in the hope of softening the increasingly violent prejudices of the Deep South against freed slaves in the post-Reconstruction era. Across the road was the Tuskegee campus, which Washington had started with next to nothing in 1881 on an abandoned 100-acre farm in Macon County, Alabama. The first classes were taught in a shanty owned by a local church. But by 1899, his new house looked out on a 2,267-acre campus, with forty-two substantial buildings, over 1,000 black students, a college-educated black faculty, and $300,000 worth of property. Just as remarkable was the fact that The Oaks, and all but four of the campus buildings, had been built by the Tuskegee students themselves, of bricks made right on campus, under the direction of a faculty member, architect Robert Robinson Taylor, one of the first black graduates of the Massachusetts Institute of Technology. Much of the carpentry, plastering, painting, plumbing, and electrical work was done by the students as practical training.

After Frederick Douglass's death in 1895, the "Age of Booker T. Washington" began. Still controversial, he was about forty-six in this turn-of-the-century photograph.

Washington had raised almost every penny for this stunning enterprise himself. And he had managed to create a viable black educational institution in hostile territory without getting himself lynched. In 1870, not many years before Washington came to Tuskegee, "nearly every colored church and schoolhouse in the area had been burned," and a few years before Washington moved into The Oaks, a lynch mob had pursued a wounded black townsman to the campus. Washington managed to defuse the situation without harm to the students or the campus, but there were more lynchings in that decade than at any other time in American history. Washington had to hire Pinkerton bodyguards when racial tensions were at their worst.

Booker T. Washington had moved into the void left by Frederick Douglass's death in 1895, and he was now the dominant black man of his time. President William McKinley had already visited Tuskegee and Queen Victoria had had him to tea. President Theodore Roosevelt would come in a few years, and later cause an uproar by having Washington to dinner at the White House. Andrew Carnegie, one of his staunchest financial supporters, considered him "the modern Moses." To historian John Hope Franklin, "[his] influence, sometimes for better and sometimes for worse, was so great that there is considerable justification in calling the period [from 1895 to 1915] 'The Age of Booker T. Washington.'"

Others felt differently. To one of his harshest black critics, Boston newspaper editor Monroe Trotter, Booker T. Washington was the "Benedict Arnold of the Negro race, the Black Boss, the Great Divider." His best-known opponent, the scholar and political activist W. E. B. Du Bois, sarcastically called him "certainly the most distinguished Southerner since Jefferson Davis."

The Oaks would be Washington's command center for the rest of his life, and legend has it that he would not die until he made it back there fifteen years later, worn out at fifty-nine from a lifetime of struggle. His second-floor den was the most important room in the house. From it, he controlled operations and interests that extended from the Alabama campus to Europe and Africa. A tireless, consummate organizer and manipula-

Washington's den was headquarters of "The Tuskegee Machine," his powerful national network of black professionals. Washington secretly directed activities on civil rights struggles that are only now becoming known. Original furnishings include gifts from admirers, such as his carved teak desk and chair.

tor, he developed what became known as "the Tuskegee Machine," a national network of black businessmen, religious leaders, educators, and other professionals he could call upon for help in pushing his own agenda, and punishing others who disagreed with it.

The house itself was spotless, staid. The Oaks epitomized the Victorian middle-class lifestyle, with all the material comforts Washington had finally achieved for himself and wanted others to aspire to. The furniture was sturdy and conventional; Washington "didn't believe in anything too elaborate," said his daughter, Portia. The walls were painted "soft and beautiful colors," she remembered, but later they became deep red, with gilded picture moldings. In the three main rooms of the first floor, an artist painted friezes of European scenes to remind Washington of his successful trip there. After dinner, Portia played her piano in the parlor for the family and guests. She'd studied music in Berlin, and was married in the parlor in 1907. Madame Schumann-Heink sang, and *Othello* was performed by members of the faculty. It's unlikely they gave the whole play because Washington was an impatient man, complaining to Portia there were just too many hallelujahs in Handel's chorus. Couldn't a few be left out to save time? Washington was a passionate gardener; the family always had fresh vegetables for dinner, and vases of flowers all over the house, from the grounds landscaped by Tuskegee's horticulturist.

Washington demanded that the campus and its students be as tidy and disciplined as The Oaks and its occupants. Portia found it difficult to live up to her father's high standards, but they adored each other. "She rests me," he would say. They liked to sit together on the porch swing in the "nice and cool" evening, and Portia was probably the only person who dared interrupt him in his den: "Papa, look at me!" she'd insist. Her mother, Fanny, who had been one of Washington's brightest students, died at twenty-six after two years of marriage. His tribute to her was revealing: "Her heart was set on making her home an object lesson for those about her who were so much in need of such help." That was the whole idea behind Tuskegee. His second wife, Olivia, mother of his two boys, died after "four happy years of married life," he wrote. "She literally wore herself out in her never ceasing efforts in behalf of the work she so dearly loved." His third wife, Margaret Murray, would become the most influential black woman of her day.

Many political and cultural notables were received in the parlor of the comfortable, immaculate house.

Like a feudal lord, Washington combed the campus on his favorite horse, Dexter, every morning he was home, noting in a little red notebook: "Steps not swept . . . pictures not hung right . . . screaking pump . . . " Missing buttons and grease spots weren't tolerated. Transgressors heard about it that very day, and made corrections or else. The faculty complained about being locked out of the dining room if they appeared one minute late.

Through his Tuskegee Machine, Washington sought to control far more than his faculty and students. He couldn't tolerate opposition anywhere from anyone of his race. He was ruthless in his efforts to destroy those who would not bend. One opponent finally decided that the only way to get away from Washington's clutches at last was to retreat to practicing dentistry.

Washington had seen enough during the Reconstruction era, he believed, to conclude that political activism wasn't the most effective way for freedmen to advance their

Off the entry hall of the Oaks were the parlor, study, dining room, and a guest room. Washington didn't like anything too elaborate, his daughter Portia said.

interests. "In many cases," he wrote in *Up from Slavery,* "it seemed to me that the ignorance of my race was being used as a tool with which to help white men into office, and there was an element in the North which wanted to punish the Southern white men by forcing the Negro into positions over the heads of the Southern whites. I felt the Negro would be the one to suffer for this in the end."

In his still controversial Atlanta Compromise speech of 1895, Washington outlined his alternative course of action. "The wisest among my race understand that the agitation of questions of social equality is the extremest folly, and that progress in the enjoyment of all the privileges that will come to us must be the result of severe and constant struggle rather than of artificial forcing." This was the essence of his difference with W. E. B. Du Bois, who believed that the demand for civil and political equality was paramount.

Washington regretted having to spend so many months each year away from The Oaks and his family, but he was raising increasingly large sums of money from Northern philanthropists and capitalists to keep Tuskegee growing, and to plug the gap that was widening between educational expenditures for whites and blacks throughout the South as traditional Southern white interests regained power. Washington was so successful that by the turn of the century, Tuskegee students studied, slept, worked, and ate in structures built with money from John D. Rockefeller, Andrew Carnegie, Julius Rosenwald, and other wealthy patrons. They admired what he was doing not only at Tuskegee but throughout the South to improve education for black students and farmers. Washington was truly a statesman in the field of education, his biographer, Louis Harlan admits.

Perhaps the most paradoxical possession in The Oaks was the Macon County, Alabama, lifetime voting certificate Washington framed and hung in his den. Washington believed that male voters—black and white—should meet certain qualifications, and he was frustrated that blacks in the South weren't doing more themselves to fight against the wholesale disenfranchisement that was occurring all over the South. But his critics were more than frustrated by Washington's own apparently weak stands. He seemed to be able to take any insult, even those delivered to his face, rationalize any injustice, see the silver lining in bigger and bigger dark clouds—even tell demeaning racial jokes to white audiences.

Washington worked relentlessly to obtain the support of wealthy and influential Americans. Two presidents visited Tuskegee—William McKinley and Theodore Roosevelt, shown here at Tuskegee in 1905.

It wasn't until his papers were studied years later that a secret side to Washington was revealed. W. E. B. Du Bois's biographer, David Levering Lewis, writes that many of Washington's rabid white Southern opponents "went to their graves never suspecting that much of the organized resistance to the extinction of the African-American as a civil being originated in the upstairs study of the Tuskegee principal." Washington had his hand in fighting and funding many battles behind the scenes: disenfranchisement cases, racial exclusion from juries, Pullman car discrimination, contract peonage.

The jury is still out on Booker T. Washington. Essentially, "his secret militancy never balanced his public conservatism," concludes biographer Harlan. Du Bois had been even harsher in his epitaph. "We must lay on the soul of this man a heavy responsibility for the consummation of Negro disenfranchisement, the decline of the Negro college and the firmer establishment of color caste." In 1967, the Reverend Martin Luther King, Jr., called Washington "a great man . . . a sincere man who misread history . . . he told us to let our buckets down where we were, and the problem was that there wasn't much water in the well."

Perhaps the jury will stay out. Washington was too complex, too ambiguous, the times in which he and Du Bois operated too volatile, too dangerous, to know if Washington should have used his prodigious talents differently even if he could have. One contemporary assessment of the epic battle between Washington and Du Bois, by the African-American economist Glenn Loury, suggests, "It cannot be said that history has proven Washington right and Du Bois wrong in their debate about what blacks should have done a century ago. Yet given the way the history of black America has evolved it can now be said that the animating *spirit* of Washington's philosophy offers a sounder guide to the future of blacks than that reflected in the world view of his critics." Others would say that it's because of those critics and their achievements that we can now better consider the question. In any event, The Oaks is a perfect manifestation of that animating spirit.

IV

FORGOTTEN FRONTIER

In this chapter there are no log cabins or ranches, no cowboys, miners, or pioneers, at least as we usually picture them. The houses here recognize the contributions of people whose presence on the frontier has been ignored or overlooked. They tell of the ancient ones, twelfth-century pueblo dwellers in the Southwest; the Cherokee Indians, reluctant pioneers forced to move west on the Trail of Tears; the Californios, the early pioneers who settled the Pacific coast long before the Bear Flag rebels and the forty-niners rushed in; and an Indian trader whose familiarity with Navajo culture helped the tribe recover from disastrous exile and survive in newer times.

The Petaluma Adobe of General Mariano Guadalupe Vallejo still appears formidable a century and a half after its construction, even though much of it has crumbled away. It is still the largest adobe ranch house ever built in California (see Petaluma, page 118).

WUKOKI PUEBLO

Wupatki National Monument, Arizona

RIGHT: *The twelfth-century pueblo Wukoki, with the San Francisco Peaks in the distance, stands on an outcropping of 200-million-year-old sandstone.*

BELOW: *Several families lived, worked, and played in this three-story, six- or seven-room house, but the site was abandoned by the early thirteenth century.*

By the time King John signed the Magna Carta in 1215, the people who had lived for about a century in the small pueblo called Wukoki, on the Colorado Plateau, had moved on. It would have been a perfect setting for the kind of epochal encounter history commemorates between the English monarch and his barons. When archaeologist Jesse Fewkes came upon Wukoki seven centuries later, in 1900, he wrote, "It is visible for many miles and from a distance resembles an old castle as it looms . . . above the plain." Wukoki stands like a fortress even today.

But Fewkes gave this prehistoric pueblo a simple name in the language of the nearby Hopi Indians, who call themselves "the peaceful people." He chose their word for "tall house." A three-story tower of red-brown sandstone still dominates what was once a six- or seven-room home for two or three families. Even as a ruin, perhaps more so without the plaster that may have once covered the meticulous stonework, Wukoki Pueblo is a dramatic architectural monument to a primal confrontation—between man and his environment.

Nearby, Fewkes found just a few clues to the identity of the families who lived at Wukoki from about A.D. 1120 to 1210: an armlet of shells, beautiful earrings inlaid with turquoise, a beautifully decorated vase, and several food bowls. Were the families who lived here members of a religious, economic, or social elite? Did they stand guard over a storehouse of food that might save them in times when even the region's annual eight inches of rain did not fall? Were they an outpost defending an ancient frontier boundary? One hundred years after Fewkes's exploration, there are no sure answers.

A yard-by-yard six-year survey in the 1980s of the Wupatki National Monument's 32,000 acres turned up evidence of four ancient cultures intermingling in the area. It appears that Wukoki was part of a great mosaic. To the Hopis of today, the inhabitants of Wukoki, like all the people who crisscrossed this

A contemporary Hopi suggests this petroglyph represents the difficulties of living in the harsh semi-desert environment and the need for spiritualty in order to survive.

desolate high desert plain for at least 11,000 years, are simply "the Hesatsinom," the ancestors, the people of long ago.

For a nation of immigrants as eager to move on as Americans have always been, it shouldn't be too surprising that the families who occupied Wukoki left home after just a few generations. They, too, could have been driven away by a prolonged drought, disease, attacks by other tribes, or even the ripple effect of the rise and fall of great population centers as far away as Mexico. No one knows. But it is certain that by the middle of the thirteenth century, Wukoki and the other memorable sites in the Wupatki National Monument were completely abandoned after a life span of about 150 years.

It amuses the Hopis today to hear people wonder, "Where did they all go?" No one seems to need to ask that question at the Acropolis in Athens. "We're still around," the Hopis say—and most of them are just fifty miles or so to the northeast, clustered around the same three mesa tops in northern Arizona where they've been for centuries. The migrations that brought the Hopis from places like Wukoki to the mesas have far deeper significance to their clans than just picking up stakes or lighting out for the Territory, as Huck Finn proposed to do. Migration as a kind of spiritual necessity is a cornerstone of Hopi clan belief: a path to understanding the meaning of life, and to an appreciation of the forces of nature. "You will go on long migrations," said the Sun Spirit's messenger, Spider Grandmother, to the people. "You will build villages and abandon them for new migrations. Wherever you stop to rest, leave your marks on the rocks and cliffs so that others will know who was there before them. . . . Never forget that you come from the Lower World for a purpose . . . compose songs to sing in your ceremonies that will remind you how the sun and moon were made, and how the people parted from one another. Only those who forget why they came to this world will lose their way. They will disappear in the wilderness and be forgotten."

It wasn't until the middle of the twentieth century that ancient Hopi myths and legends began to be told to outsiders for the written record. Some idea of their historical and religious worldview began to be understood. One listener felt he was hearing a "cosmic drama." To another it was as if an epic worthy of Homer were being told, with its "great sweep of human conflicts and supernatural events."

Wukoki was one of the last stops along the way to what the Hopis believe is the "Center of the Universe," the mesa-top villages they now inhabit. This is where they settled after they had completed the migrations the god of the Upper World, Masauwu, commanded them to take–to the north, south, east, west and back–when they first emerged into it. This is where their struggle between good and evil continues. The Hopi people know they are being tested once again in this dry, desolate place. They must not lose faith in their Creator, who brings the rain that gives them life.

To some Hopis, our present world, the Fourth World, appears to be on its way to destruction. Emergence to the Fifth World has already begun. The world will end this time, one Hopi believes, unless all human beings begin to cooperate with one another to save the human race, and the planet. Another observer sees a war between the spiritual and the material as the final struggle.

So Wukoki is part of a far larger, longer Hopi history. But there is something very special about this pueblo, small though it is. Those who prefer it to the world-famous pueblos and cliff dwellings elsewhere in the Southwest see it as a place of great spirituality, an intensely romantic spot. It casts a powerful, lasting spell.

Maybe it's the delicate beauty of the pueblo itself, so skillfully constructed from slender slabs of sandstone it seems to

ABOVE: *Edward Curtis photographed these Hopi maidens in 1906, and there are still strong links between the Wukoki site and contemporary Hopi villages not far away.*

BELOW: *The graceful stonework is made of Moenkopi sandstone slabs and was once probably covered with plaster.*

rise up effortlessly out of an outcropping of the same rock laid down by a shallow sea 200 million years ago. Perhaps it's because there isn't another natural or man-made structure within sight for miles around, nothing to distract attention from the almost palpable presence of those who lived here centuries ago. The air is so clear, the absence of mechanized sounds so complete, you can almost hear the echoes of an ancient family's daily life in the sheltered plaza beneath the high tower.

It could be the unsettling recognition that the people who lived in this harsh place could count on an ancient and encyclopedic cultural heritage of survival skills that modern man has lost. They sited their pueblo so that the plaza received direct sun; the sandstone slabs retained heat for winter warmth, and provided insulation from summer sun. Even though each person living at Wukoki probably needed less than two gallons of water a day to survive, where did they find it? There's no known present-day source of water within two and a half miles. How did they grow the food they ate, the cotton they wove and traded? Hopi techniques for diverting or containing every precious drop of water, and planting corn seed a foot deep to take advantage of what moisture there is in the soil, have been practiced for centuries and provide clues.

It takes just a brief stay at Wukoki to begin to appreciate the origins of ancient Hopi beliefs. Thunderclouds billow every summer afternoon from the snowcapped San Francisco Peaks south of the pueblo, 2-million-year-old volcanic mountains as sacred to the Hopis as, say, Mount Olympus was to the ancient Greeks. The clouds, and the rains they bring, might very well be the kachinas, the spirits of the ancestors, who live there part of the year—before they are welcomed back to the Hopi villages in midwinter to share their lessons about life, and how to live it virtuously, so that rain will continue to come. And who can say that the spirit of the wind, Yaponcha, doesn't still live in a deep crack at the base of Sunset Crater?

Not long ago, a tribal elder said, "We Hopis are the last of the frontier. Everything else of the American frontier has changed except the Hopis, who still do the traditional things more completely than anyone else. . . . We have survived because it was in our prophecies."

Hopi artists of our time draw their inspiration from these traditions and prophecies, and from the physical beauty and spiritual power that awe visitors at Wukoki. Ramona Sakiestewa, the weaver, uses a twelfth-century textile pattern found not too far away to create a magnificent blanket design she has named "Wupatki." This mysterious fabric embodies the site, she explains, and the people who survived here centuries before. Hopi painter and sculptor Dan Namingha, a descendant of the Hopi potter Nampeyo, draws on the landscape, the kachina spirits, and ancient migration symbols etched in the surrounding rocks for his striking abstract paintings that are now part of public and private collections around the world. "My foundation always is my culture," he says. "I am a part of something that goes centuries back."

Hopi poet Ramson Lomatewama may have found the clearest contemporary meaning for all Americans in a place like Wukoki, where the overpowering ancient landscape helped shape the Hopi view of life as a spiritual quest toward harmony between man and nature.

This blanket is one of six in the new Ancient Blanket series by contemporary Hopi weaver Ramona Sakiestewa. The Wupatki pattern is taken from an ancient textile found in the vicinity of Wukoki.

Cloud Brothers

Four directions
cloud brothers
share one sky.

Each has its own path.

Each has its own mood.

Each has its own face.

The cloud brothers are many
but they are one family.

The cloud brothers are scattered
but they are one spirit.

They mingle
within themselves

changing with every moment.

They tell us
that we too
are brothers
on this land.

And

like our cloud brothers

we are all yellow
as are the sunrise clouds

we are all white
as are the noonday clouds

we are all black
as are the thunder clouds

we are all red
as are the sunset clouds.

So let us look up to our cloud brothers
as one family
and one spirit.

For we are truly different

and yet

we are truly the same.

CHEROKEE CHIEF VANN'S HOUSE

CHATSWORTH, GEORGIA

WHEN JENNY LIND, THE SWEDISH NIGHTINGALE, SANG "HOME, Sweet Home" at a concert in Washington in 1850, President Millard Fillmore and orator Daniel Webster were in the distinguished audience. Her rendition reduced Webster to tears, it was reported. The composer, John Howard Payne, was also in the audience, but his reaction was not noted. The song was world-famous by then, and legend had it that Payne wrote the words on the basement walls of Cherokee chief James Vann's house, where the Georgia Guard was holding him prisoner for inciting the Cherokee Nation to protest attempts to remove them from the territory they thought had been theirs for centuries.

This brick mansion was a showplace of the Cherokee Nation at a critical time in the tribe's history. It was built by slaves and Indians for Cherokee town chief James Vann.

Part of the story is accurate. Payne was imprisoned for several weeks in the fall of 1835 on Chief Vann's property, and he had become an eloquent spokesman for the Cherokees in their eleventh-hour struggles to stay on their lands. But his jail was a log hut on what had been a prosperous plantation in northwest Georgia, part of an extensive, diversified commercial operation run by Vann, a Cherokee town chief, and later by his son Chief Joseph Vann. And whatever support Payne had been able to create nationally for the Cherokee's cause came too late. So if Payne wept any tears at Jenny Lind's concert, it was for the painful memory of the writing on the wall for the Cherokee Nation. In 1838, just three years after his release from jail, over 16,000 Cherokee men, women, and children were forcibly moved on "the Trail of Tears" to Indian Territory west of the Mississippi. Four thousand perished along the way.

Chief James Vann's house was a brilliant but futile showplace for the Cherokee Nation's impressive achievements. Son of a Scot trader and a Cherokee chieftain's daughter, he was one of the mixed-bloods in the Cherokee Nation who prospered to help the tribe financially and politically in its remarkably successful efforts to become "civilized," as the new American government wished. Even before independence, the Continental Congress decided "a friendly commerce between the people of the United Colonies and the propagation of the gospel and the cultivation of the civil arts among the [Indian tribes] may produce many and inestimable advantages to both." In 1802, Thomas Jefferson went further:

> We shall with great pleasure see your people become disposed to cultivate the earth, to raise herds of useful animals and to spin and weave, for their food and clothing. These resources are certain, they will never disppoint you, while those of hunting may fail, and expose your women and chil-

> dren to the miseries of hunger and cold. We will with pleasure furnish you with implements for the most necessary arts, and with persons who may instruct how to make and use them.

James Vann had taken full advantage of everything that came his way, but he survived only four years in the house he moved into on March 24, 1805, according to the daily diary of the Moravian missionaries he'd invited to set up a school nearby on his land. When Vann was shot dead in 1808, in revenge for killing his brother-in-law in a duel, their epitaph was as extraordinary as Vann himself.

> Thus ended the life of one who was feared by many and loved by few in the 41st year of his life. . . . Vann had been an instrument in the hand of God for establishing our mission in this nation. Never in his wildest orgies had he attempted to harm us. We could not but commend his soul to God's mercy.

Composer Payne later allowed that "[Vann] with all his errors . . . was a patriot."

The missionaries had helped Vann build his house, but they seemed to be the only people who weren't impressed. All the materials except the windowpanes came from Vann's own operations. His slaves and Indians made the bricks and carved the woodwork. He may even have had a German architect named Vogt. The result was a handsome Federal-style mansion, with Georgian porticœs front and back, the first brick house in the Cherokee Nation.

The residence was so imposing that President James Monroe chose to stay there rather than at the Moravian mission during a four-month swing through the South in 1819. His host was James's twenty-one-year-old, six-foot-six son Joseph, who had taken over his late father's business affairs, and was even more successful. By the time Joseph was driven out of Georgia in the winter of 1834–35, he was one of the two richest men in the Cherokee Nation. But "Rich Joe's" wealth could not save him. Even as the Vann house was going up in 1805, the missionaries had foreseen, "The whites are gradually encroaching until the Cherokee will finally be completely crowded off their land."

By the time President Monroe visited the Vann house, voluntary relocation of the Native Americans remaining east of the Mississippi was believed to be the best solution, in the white man's eyes. Separation, not assimilation through civilization, had become the most practical policy, at least to the government in Washington. Already, the tribes of the old Northwest Territory had moved across the Mississippi, out of the way of the pioneers pushing westward, and some Cherokees, amid bitter and bloody debate, had decided to follow. Monroe told Congress:

OPPOSITE TOP:
The woodwork throughout the house is painted brilliant green, red, yellow, and blue, colors representing Cherokee forests, soil, grain, and sky. A portrait of James's son, Chief Joseph Vann, hangs over the dining room fireplace.

OPPOSITE BOTTOM:
The twelve-foot mantelpiece in the drawing room shows the green ceiling and yellow "Cherokee rose" motif that appear throughout the house. A two-hundred-year-old Cherokee double-weave basket is on the hearth.

> It was right that the hunter should yield to the farmer, for the earth was given to mankind to support the greatest number of which it is capable, and no tribe or people have a right to withhold from the wants of others more than is necessary for their own support and comfort . . . but forced removal would be "revolting to humanity, and utterly unjustifiable."

The president slept in the well-appointed guest room on the second floor. James Vann had seen the best in his many travels, and the house reflected it. His showplace was about as civilized as you could get in the backwoods of Georgia. Son Joseph was well traveled, too, although the missionaries despaired when James took his son on business trips; he was enough of a problem, they thought, without exposure to his father's wild ways.

The guest room was one of four spacious rooms in the house on the first and second floors. Four large-paned glass windows in each room looked out on Vann's thriving estate. The "coffin rooms" in the attic accommodated children, servants, and travelers—probably on straw pallets. The 12-foot ceilings in the guest room, the master bedroom, the dining room, and the drawing room were of wood painted bright green, one of the four vivid colors used throughout the house: on the wainscoting, the moldings, and the mantelpieces with hand-carved Cherokee roses, scorpions, lizards, snakes, and frogs. The astonishing floating staircase in the wide hallway between the drawing and dining room leading to the upper floor was all green, too, with yellow Cherokee roses. The green represented the trees and grass on the Cherokee Nation's 15 million acres; the yellow, ripened grain; the blue, Cherokee sky; the reddish brown, the Georgia clay soil. President Monroe must have noticed the "Christian doors" in the house, perhaps a Moravian touch. Each had a cross

motif on the top half, and a representation of an open Bible on the lower half in various color combinations.

The Cherokees weren't interested in adopting everything the white man offered, religion or otherwise. They picked and chose. "They did not need the example of the American Constitution to understand the value of check and balances," one scholar points out. But they did value white laws and institutions that protected the individual. They could also count the figures in their own 1824 census, which showed they were materially much better off than they'd been thirty years before. "I am sensible the hunting life is not to be depended on," said Major Ridge, one of the Cherokees' most capable leaders. His son, John Ridge, educated by the Moravians, was even more pointed, speaking at a church in Charleston in 1823:

> Will anyone believe that an Indian with his bow and quiver, who walks solitary in the mountains, exposed to cold and hunger, or the attacks of wild beasts, trembling at every unusual object, his fancy filled with agitating fears, lest the next step should introduce his foot to the fangs of the direful snake . . . actually possesses undisturbed contentment superior to a learned gentleman of this commercial city, who has every possible comfort at home?

Both men later paid with their lives for deciding removal was the only option for the Cherokees to save themselves.

A Moravian missionary observed wryly in his diary in 1802, "They were more interested in the three Rs than the Trinity." James had sent Joseph to the Moravian mission school, Joseph sent some of his own children, and several key Cherokee leaders had been educated there, too. But neither James nor Joseph was among the pitifully small number of Cherokees converted after more than a generation of backbreaking effort. James did have a copy of the New Testament (and *The Young Man's Best Companion).* He admitted to the missionaries that "he knew he ought to read it, but always neglected it."

It's hard to see what more the Cherokees could have done to become "civilized." In Philadelphia, John Ridge argued:

> You asked us to throw off the hunter and warrior state: We did so–you asked us to form a republican government: We did so–adopting your own as a model. You asked us to cultivate the earth, and learn the mechanic arts: We did so. You asked us to learn to read: We did so. You asked us to cast away our idols, and worship your God: We did so.

President James Monroe may have stayed in this second-floor guest bedroom on his Southern tour in 1819.

Even more remarkable, one of their own, Sequoyah, who could not speak, read, or write English, created a Cherokee alphabet, making Cherokee the first Native American language to be written and read. They published their own newspaper, the *Phoenix*, the first Native American newspaper. They adopted their own constitution and built their own Supreme Court building. This poignant structure has been re-created on the site of their ill-fated capital, New Echota, not many miles from Vann's house.

But it appeared difficult, even impossible, for many whites to recognize what the Cherokees had achieved. The Moravians noted in their August 1829 diary that even the governor of Tennessee had no idea the Cherokees had made such progress toward civilization: "Until now he had thought them wild people who lived in miserable huts." President Andrew Jackson's secretary of war, John Eaton, felt "the Indians were no more

ABOVE: *The Cherokees were removed forcibly to Oklahoma in 1838. Their "Trail of Tears" is depicted in this famous painting by Robert Lindneux. The tribe's successful efforts to become "civilized" were disregarded, and 4,000 Cherokees perished on the way west.*

BELOW: *This buckskin coat, known as the "Trail of Tears" coat, is said to have been worn during that forced march. It is decorated with ribbon work and a bear paw design in the lining.*

capable of being educated than wild turkeys." By the time the U.S. Supreme Court decided several cases in the Cherokees' favor, in 1830 and 1832, it was too late to change attitudes or actions. President Jackson was an advocate of *forced* removal. "John Marshall has made his decision, now let him enforce it," Jackson is quoted as saying. A Cherokee warrior, Chief Junaluska, regretted saving Jackson's life during the Creek War. "If I had known that Jackson was going to drive us from our homes, I would have killed him that day at the Horseshoe."

In the fall of 1833, Joseph Vann made a mistake that cost him everything, including his showcase house. For years, Georgia had been squeezing the Cherokees in every possible way. The tribe's successes had made their properties too desirable, and when gold was found on Cherokee land, the end was near. Before Joseph left on a business trip, he'd hired a white man to oversee his estate for the year 1834. Vann didn't realize it had become illegal in Georgia for a Cherokee to do that. Vann let the man go as soon as he found out, before the new year began, but this was just the excuse Colonel William Bishop of the Georgia Guard was waiting for to get Vann out of his mansion for good. Gunfire didn't drive Vann out, so Bishop tried something else. On the landing of the impressive cantilevered hall staircase, burn marks from the firebrand Bishop threw in to smoke out Vann and his family are still visible. That worked. The Vanns fled first to Tennessee, then to Oklahoma, where Joseph built up a successful business again, even constructing a replica of his Georgia mansion. It was destroyed during the Civil War.

This shoot-out wasn't the only skirmish Bishop fought at Vann's house. Two years later, the Moravian missionaries were also on their way out of Georgia. Bishop had rigged an election, and his furious opponents marched on the house in protest. One of the last missionaries to leave wrote:

> Colonel Bishop [is] forted—in the brick house (Vann's formerly) & from sundry windows & extra port holes, pro-

> jected the ghastly muzzles of muskets & Rifles—threatening death & destruction to all who should possess the bold daring to attempt a reduction of the Castle. . . .

Joe Vann met his own violent death at age forty-six in a boiler explosion that tore his steamboat apart during a reckless race on the Ohio River. The Moravians had always been scandalized by Rich Joe's horse races too close to their mission, and the boat was named the *Lucy Walker* for one of his favorite horses. Vann's body was never found. His ring, with a picture of "The Lucy Walker" engraved on it, and his violin are the only original Vann possessions that survive in the house. The Moravians didn't write an epitaph for Joseph Vann. Those who might have composed one for the Cherokee Nation would have been premature. Today more Native Americans identify themselves as Cherokees than as members of any other tribe. What did die was an extraordinary opportunity for the new nation to work with a talented, flexible, and far-seeing people on a solution that "might have created a framework for tolerance and ethnic diversity that would have spared the nation much of the racial conflict that has tarnished its political landscape ever since." The Vann House is a poignant reminder of that lost opportunity.

ABOVE LEFT: *Burn marks on the staircase landing remain from the firebrand thrown into the house in a successful effort to drive Chief Joseph Vann and his family out of the territory, a campaign that intensified when gold was found on Cherokee land.*

ABOVE: *The front door beyond the "floating" or "hanging" staircase may show nearby Moravian missionary influence in its cross and open Bible design.*

GENERAL MARIANO GUADALUPE VALLEJO'S PETALUMA ADOBE

PETALUMA, CALIFORNIA

"IT WAS A CASE OF LOVE AT FIRST SIGHT," the young army officer wrote his superiors in the spring of 1833. The brilliant, ambitious Mariano Guadalupe Vallejo, twenty-five years old and already *comandante* of the San Francisco Presidio, was passionate about many things: his family's honor; his beautiful new wife, eighteen-year-old Doña Francisca Benicia Carillo, and the promise of a prosperous future for California, tied in some way to the new American Republic to the east, which he admired greatly. Here Vallejo was rhapsodizing about the strategic Petaluma Valley, which he'd passed through on a mission to Mexico's northern frontier to find out what Russian settlers at Fort Ross were up to. They had moved too close for comfort—less than a hundred miles from San Francisco. They had to be stopped, and if anyone could do that, it was Mariano Guadalupe Vallejo.

The huge courtyard of the Petaluma Adobe was the center of an agricultural and industrial enterprise involving 2,000 workers.

"The great valley filled me with emotion," Vallejo went on, with typical exuberance. "Nowhere was there a scene of such beauty and the suggestion of everything desirable for man. . . ." Less than a year later, the valley was his, part of a huge 100-square-mile grant of land from the Mexican government in lieu of his long-overdue salary. By 1846, when California was taken over by the Americans, Vallejo held title, or so he thought, to 175,000 acres, and he was one of the wealthiest and most powerful men in Alta California, with a reputation that reached Europe and assets close to a million dollars, by his own estimate. The fortresslike adobe hacienda he began to build in April 1836 on a knoll overlooking Petaluma Valley was massive, deliberately intimidating. Far-seeing though Vallejo was, he could not have predicted that it would become a poignant symbol of his own rapid rise and faster fall, and the end of a pastoral, paternalistic way of life cherished by the Californios, those native-born men and women like Vallejo and his wife, whose distinguished roots in the New World went back to the sixteenth century. "We were the pioneers of the Pacific Coast," Vallejo would bitterly remind the

Americans later, "building pueblos and missions while General Washington was carrying on the War of Revolution." Vallejo might have added that he had successfully subdued the Indians in his territory.

By the time Vallejo was twenty-eight, he had been appointed military commandant and director of colonization for Mexico's northern frontier. The Petaluma Adobe was now the headquarters of an enormous agricultural and industrial empire with thousands of workers. At its peak, the enterprise brought Vallejo $96,000 a year from trade in hides and tallow alone. The Petaluma Adobe wasn't Vallejo's only residence, but it was by far the biggest; it is still the largest privately owned adobe structure ever built in California. The four walls formed a huge quadrangle at least 200 by 145 feet, enclosing an enormous open-air courtyard where Vallejo's laborers assembled for roll call every morning. The walls were three feet thick, made of large bricks fashioned from the black soil on the site, which added to its menacing appearance. The sight of the place looming over the valley sends shudders through an approaching visitor even today, when half of it has melted away. There was just one entrance into the courtyard, and the windows were barred against Indian or Russian attacks, which never came. The Petaluma Adobe would be overcome in a decade by forces Vallejo did not anticipate: the uncontrollable onslaught of immigrants from the East.

Vallejo was a prolific, colorful writer. His letters to Doña Francisca were lusty even after years of marriage and sixteen children; his voluminous history of California is still significant. Vallejo was proud of saying, "My biography is the history of California." His description of life at the Petaluma Adobe was just as vivid:

> . . . The house was of immense proportions, owing to the different departments for factories and warehouses. I made blankets enough to supply over 2,000 Indians; also carpets and a course *[sic]* material used by them for their wearing apparel. A large tannery also, where we manufactured shoes for the troops and Vaqueros. Also a blacksmith shop for making saddles, bridles, spurs, and many other things required by the horsemen. . . . My harvest productions were so large that my storehouses were literally overfilled every year. . . . All these products were stored in different departments of this large house which also supplied the Indians who lived around the surrounding country and in peace with me. A large number of hides were preserved every year, also tallow, lard and dried meat to sell to the "Yankees."
>
> In one wing of my house, upstairs, I lived with my family when in Petaluma. . . .

ABOVE: *The kitchen at the Petaluma Adobe reflects the size of Vallejo's operations there.*

LEFT: *The Vallejo family's suite was luxurious by frontier standards. It was the only part of the immense structure with glass windows.*

> The house was two stories high and very solid, made of adobe and timber, brought from the redwoods and cut for use by the old fashioned saw by four servants brought from the Sandwich Islands. . . . It had broad corridors, some of which were carpeted with our own made carpets. . . .

The Vallejo family quarters may have seemed simple to visitors unaccustomed to frontier life, but their rooms were spacious, with carpets, fireplaces, smooth walls, finished ceilings, and windows of costly imported glass. The General, as he came to be called, and Doña Francisca, his "frank, gay and engagingly provocative" wife, as he described her, came to the rancho a few weeks each year on holiday, with their children, relatives, and servants for every conceivable job (five just to grind corn for the tortillas). The heat from the large ovens in the courtyard, the noise from the blacksmith's bellows, the carpenter's shop, and the slaughtering of cattle; together with the smells from drying hides, boiling tallow, rotting meat, and human sweat all would have been overpowering for a much longer stay. The Vallejos' arrival meant fiestas, fandangos, huge roundups of some of his thousands of head of cattle, even fights between bulls and bears in the courtyard. The family spent the rest of the year at La Casa Grande, another imposing adobe he built in nearby Sonoma, the town he founded, and where he garrisoned his troops. La Casa Grande was more suitable to Vallejo's growing reputation as "the Lion of the North," the unofficial host of California. It was immaculate, comfortable, and handsomely furnished.

La Casa Grande became a must stop for visiting dignitaries from the East and Europe, who came to cultivate Vallejo's good will and support for their particular national or individual interests. Doña Francisca fit in well at either house. "Voluptuous, contented and happy" wrote one admiring visitor, "[she] possess[ed] in the highest degree the natural grace, ease and warmth of manners which render the Spanish ladies so attractive and fascinating to the stranger," said another. She liked the Paris gowns and $600 tortoiseshell combs Vallejo could afford to buy her in his heyday. But she was very much a woman of the frontier, level-headed during any crisis, a skilled markswoman who could deliver babies and baptize them officially for the church when she took charge of the estate in her husband's absences. To General Vallejo, she had the most beautiful face he had ever seen.

The Vallejos' generosity and hospitality were legendary. William Boggs, the twenty-year-old son of one exhausted family that had come overland from Missouri in the winter of 1846–47, described their welcome at Petaluma:

> On our arrival in the night at the ranch, General Vallejo, who had gone ahead of our worn-out teams, had aroused his Indian servants to prepare supper for us. The tables were spread with linen table clothes; sperm candles were in the chandeliers, and we had a regular Spanish cooked meal, wholesome and plenty of it. With Spanish hospitality, the General waited on the table, helping all the large family. After supper he handed Mrs. Boggs a bunch of keys to the various rooms and assigned one well-furnished apartment to her and me.

General Vallejo's hospitality was legendary. He served dinner himself in the adobe's dining room to an exhausted immigrant family just arrived overland from Missouri.

Mrs. Boggs had a baby boy there just a few weeks later; Vallejo was always proud to claim Guadalupe Vallejo Boggs as the first child born in California of American parents. The Boggs family stayed at Petaluma all that winter. In 1892, his namesake wrote:

> The General never met me in the street or in any place but that he had to pat me on the head and tell me when I was an infant, he and his wife saved my life by having a sheep killed and wrapped me in the warm pelt to keep vitality in me.

Vallejo greeted other American immigrants as cordially, if he thought they could contribute something positive to his beloved California. In 1841, he'd admitted the first overland wagon train of American immigrants, led by John Bidwell. Later, Vallejo gave 320 acres of his land to a Methodist minister, Lorenzo Waugh, who remembered Vallejo's welcome:

> You are the kind of man we want; families come to stay, to make their homes, and to cultivate our beautiful lands . . . land is plenty. . . . God made it for us, and I have plenty of it yet.

Vallejo could afford to give 3,000 acres to his children's music teacher in exchange for piano lessons, and 11,000 acres to the North Carolina–born trapper, George Yount, who had demonstrated the advantages of redwood shingles over thatch for the roof of Petaluma Adobe. Vallejo was always open to new ideas, new technologies.

Vallejo realized that some relationship to the United States was inevitable. "We belong to the American continent and are opposed to European heads," he told foreign visitors. The family biographer credits Vallejo with being "the first person on the Pacific Coast to endorse the principle of the Monroe Doctrine—that no European power could be permitted to extend its system over any government in the Western Hemisphere."

But on June 14, 1846, that relationship took a devastating turn for Vallejo and the Californios. "A little before dawn," he wrote later, "a party of hunters and trappers with some foreign settlers . . . surrounded my residence at Sonoma and without firing a shot made [me] a prisoner." "To what do I owe the visit of so many exalted personages?" the proud Vallejo asked sarcastically as they invaded his house. For the next six weeks, Vallejo, his brother Salvador, his American brother-in-law, and his private secretary were held captive at Captain John Sutter's fort in Sacramento. Conditions were frightening ("We have not been killed, at least up to this time") and humiliating ("And they tied me to a chair! Me! Vallejo!"). If he'd known that his wife had been threatened with execution by one of the leaders if she tried to smuggle out arms, he would have been frantic. By the time Vallejo was released in early August, he was emaciated, ill with malaria, and angry. He'd been held hostage by a small group of rough, tough, well-armed Americans (even John Sutter called them "ruffians," the tamest epithet of many), acting without the approval of the U.S. government at the instigation of the controversial Captain John C. Frémont. No one is sure even now, but Frémont may have wanted to remove Vallejo as a future political rival. Vallejo's brother Salvador wrote later:

> My heart grieved for my brother. . . . I thought of the many English, Americans, French, and Russian officers, that had received kind treatment at his hands. . . . And when the light of day allowed me to see him lying on the damp floor without coverings or even a pillow on which to rest his head, I cursed the days in which our house dispensed hospitality to a race of men deaf to the call of gratitude, so perfect strangers to good breeding.

General Vallejo, whose ancestors came to the New World in 1500, became one of the most powerful and prosperous men in California under Mexican rule.

The American flag had been raised over California while Vallejo was imprisoned. Señora Vallejo went to the "door every half hour to be sure no daring hand would attempt to remove that flag on which she said hung all her hopes," wrote one observer. Some of Vallejo's countrymen felt at first that cocky, autocratic Vallejo got what he deserved for being so welcoming to the gringos. Salvador Vallejo, always cynical and suspicious of American motives, had berated Vallejo for years. So had Don José de Jésus, an older brother. None of them knew it then, but this "Bear Flag Rebellion" was the beginning of the end for all of them.

"I left Sacramento half dead," Vallejo wrote later. "The political change has cost a great deal to my person and mind and likewise to my property. I have lost more than a thousand live horned cattle, six hundred tame horses, and many other things of value which were taken from my house here and at Petaluma. My wheat crops are entirely lost. . . . All is lost, and the only hope for making it up is to work again." Vallejo tried—over and over. But Petaluma and Vallejo's fortunes never recovered from the Bear Flaggers' rampage and the flood of immigrants from the East that followed. The early pioneers, the Californios, even Vallejo, with his influence, wealth, and admiration for "the American ideas, theory and practice of vigorous growth and improvement of California," never had a chance. By the end of 1849, the year of the Gold Rush, close to 100,000 new pioneers had arrived. The Californios, including the "pioneer of pioneers," Vallejo, were overwhelmed physically, culturally, and politically, as they had overwhelmed the Indians.

Richard Henry Dana, Jr., the Boston brahmin turned sailor, author of the 1840 bestseller *Two Years Before the Mast,* summed up the often contemptuous American attitude: "In the hands of an enterprising people, what a country this might be!" American trapper James Clyman wrote that the Californios were "a proud, indolent people doing nothing but ride after herds from place to place without any apparent object." Clearly, the old Californios weren't worthy of California, in their view. Their work ethic was too different. So was their religion, their family structure, even their color.

The Petaluma Adobe had deteriorated badly by 1880, after the Californios, early pioneers like Vallejo, were overwhelmed by the new pioneers from the East.

To the Californios, on the other hand, "the Anglo-Saxon's preoccupation with labor, profit and savings for the future always remained something of a mystery." Vallejo offered an ironic version of this profound culture clash to President Lincoln:

> The Yankees are a wonderful people—wonderful! Where ever they go, they make improvements. If they were to emigrate in large numbers to hell itself they would irrigate it, plant trees and flower gardens, build reservoirs and fountains and make everything beautiful and pleasant, so that by the time we get there, we can sit at a marble-top table and eat ice cream.

Vallejo had figured his land titles were protected by the Treaty of Guadalupe-Hidalgo, which transferred half of Mexico to the United States in 1848 for $15 million. The conqueror "seeks his own good fortune, not ours," he finally realized. This is "very natural in individuals, but I denounce it on the part of a government that promised to respect our rights and treat us as its own sons." Vallejo spent years, and thousands of dollars, on lawyers (he called them a "swollen torrent of shysters" from the States), trying to prove the validity of his titles to his lands.

The Petaluma Adobe, built for an estimated $80,000, had gone for a song, $25,000, in 1857, to pay legal fees. Vallejo's reliance on the validity of his title to the vast Soscol rancho was shattered by a Supreme Court ruling in March 1862; La Casa Grande burned to the ground in 1867. By the time he died, all Vallejo had left was 228 acres, one cow, two aging horses, and a small Carpenter Gothic house with the appropriate name Lachryma Montis (Mountain Tear). To his old friend Lieutenant Joseph Warren Revere, Paul Revere's grandson, this was the "most shameless and barefaced robbery." But Vallejo told his son, Dr. Platon Vallejo, the first native-born California physician, "I brought this on myself. . . . It was best for the country, and . . . I can stand it."

In 1880, Vallejo reminisced about the old Petaluma ranch house:

> It is a sad memory but one bows to that which says that "all is perishable in this world." I compare that old relic with myself and the comparison is an exact one; ruins and dilapidation. What a difference between then and now. Then Youth, strength and riches; now age, weakness and poverty.

Platon shared his nostalgia:

> [In my dreams] I would see myself a little child . . . looking down and away into the beautiful valley covered with flowers and filled with beautiful cattle and horses rolling in fat and happily sunning themselves in the sun's golden light like a blessed people enjoying existence in a happy land. Petaluma has a charm for me and there is an enchantment about that place even yet. . . .

For a century since, Americans have gone to California in search of such enchantment. Reverend Walter Colton, a Vallejo admirer who'd also fallen in love with the Californios' way of life, was quite clear where he thought it was to be found:

> If I must be cast in sickness or destitution on the care of the stranger–let it be in California, but let it be before American avarice had hardened the heart and made a god of gold. There is hardly a shanty among [the Californios] which does not contain more true contentment, more genuine gladness of the heart than you will meet with in the most princely palace.

Vallejo was always full of life; even in his last year he'd get out of bed to do an Indian war dance. He knew "the Wheel of Fortune is very fickle" " . . . and you could not fly with heavy cargo to the celestial regions." None of his or his wife's entrepreneurial schemes after the American takeover came close to making up the fortune he'd lost: not his disastrous offer of land and buildings for a new state capital; not the unsuccessful offer of the Petaluma Adobe to the University of California for its first campus; not the quarry that provided the blocks to pave San Francisco's streets; not even the award-winning vineyards and orchards that to one admirer make Vallejo the father of horticulture in California. But Vallejo and other Californios down to the lowliest miner and vaquero, who passed on critical new skills to their American counterparts, left much behind for America. "If Vallejo was not actually the founder of California's diversity," one biographer concludes, "he was certainly one of its chief architects. . . . His faith in reason, education and progress led him to believe that the United States might one day prove to be the hope of the world." Perhaps it is this persistent vision of the best of America, even though he had also seen its worst, that is Vallejo's greatest legacy.

LORENZO HUBBELL'S HOUSE

HUBBELL TRADING POST
GANADO, ARIZONA

RIGHT: *The Hubbell House, with the white porch, is in the foreground in this view of the Hubbell Trading Post complex in Ganado, Arizona.*

BELOW: *John Lorenzo Hubbell's initials appear on the gate to his house, directly behind the trading post.*

AT THE TURN OF THE TWENTIETH CENTURY, JOHN Lorenzo Hubbell ("Old Mexican" to the Navajos) built a new adobe house behind the trading post he'd been operating for twenty-five years on the Navajo Reservation in eastern Arizona. Like everything else Hubbell put his enormous energy and initiative to, his house was massive and solid. Together with the dimly lit, fully stocked trading post next door, the stone barn, corrals, smithy, and bakery, it was the hub of a trading empire Hubbell had begun to build in 1876, when he was in his early twenties.

By 1900, Hubbell ("Don Lorenzo" to the Anglos and Hispanics) was well on his way to becoming a legend among Indian traders. His personal courage had been well tested, deliberately by the Navajos and dangerously in the range war between Texas cattlemen and New Mexican sheepmen while he was sheriff of Apache County. His tales of bullet wounds, narrow escapes, and dramatic rescues were engrossing, if tall. Hubbell's door was always open; so was his pocketbook. "Ganado was a place of refuge," said a Navajo woman. "Friends don't just take off, but stay there to the end. You don't leave someone behind. That's what Ganado means to me." A white friend called him "the most hospitable man in the world. It was nothing for him to entertain one hundred and fifty people at the post at a time—Indians, Mexicans, bullwhackers, Eastern tourists, anthropologists, archeologists, ethnologists." "I never charged anyone for a meal or a night's lodging," said Hubbell, who may have been the trader who began the trading post custom of offering snacks and tobacco before getting down to business. Many of his Navajo customers, interviewed years later, remembered that gesture vividly; it was their custom as well.

Trading with the Indians was nothing new in the Southwest or anywhere else in North America. But the licensed trading post on the reservations of the Southwest, run by men of probity subject to regulations on profit margins and what they could or couldn't sell,

Hubbell (with mustache, on the left in the wareroom) was one of the most influential, reputable traders on the frontier.

was a post–Civil War development. Since George Washington's day, the United States government had attempted to regulate trade with the Indians to prevent abuses. But its inability to control all traders led to the stereotype of the Indian trader as "a first class scoundrel—a man who attains financial prosperity by fleecing the Indians," as Lorenzo Hubbell himself put it. He wasn't "a money-grabbing, gun and whiskey-selling rascal" like some. Lorenzo Hubbell was one of the best. His reputation for honesty and wise counsel to the Navajos has held up under hostile scrutiny ever since.

Hubbell came to know Navajo territory well. He was the ideal person, in an ideal position at Ganado, to act as a bridge between alien cultures: the Navajo, the Hispanic, and the Anglo. Hubbell had been born and raised in New Mexico. His father, a New England Yankee, had come to New Mexico as a soldier. Hubbell boasted that his mother's parents were descended from old Spanish-American families who had lived in that part of the country since the time of the conquistadores. Hubbell spoke Spanish before he spoke English, and he had picked up Navajo on the odd jobs he had as a teenager.

Years later, one of the Navajo women who traded at Hubbell's Ganado post criticized his Navajo; to her it sounded like baby talk. But it was good enough for him to serve as interpreter in some tense disputes between the Navajos and their Anglo and Hispanic neighbors. Traders couldn't get by without speaking a little, but its complexities were beyond most, and few stayed around as long as Hubbell did. It was a tough, lonely, and not very profitable business.

Hubbell's career as one of the best-known and best-liked Indian traders spanned what was a critical half century for the Navajos. Indian traders like Hubbell helped them adjust to, and exploit if they could, their new political and economic status as a conquered people. And it was traders like Hubbell who helped the white world appreciate and understand Navajo culture and its strengths. "Neither the trader nor the Indian could have survived economically without the other . . . it was a symbiotic relationship," says anthropologist Edward T. Hall.

"Out here in this country," Hubbell explained in a last interview before his death in 1930, "the Indian trader is everything from merchant to father confessor, justice of the peace, judge, jury, court of appeals, chief medicine man and *de facto* czar of the domain over which he presides." He could have added a few more roles: funeral director (Navajos relied on non-Navajos to facilitate burials); employment agent (traders handled the hiring of Navajos for work off the reservation); impresario (traders helped pay the costs of major Navajo ceremonies, and held "chicken pulls," the Navajo version of a rodeo, as close to their posts as possible, to attract customers and tourists); and "son of the Great White Father in Washington" (the government relied on traders for all kinds of help to get Navajo support for their policies and programs).

Oddly enough, Hubbell's list left out perhaps his most tangible and lasting contributions. A visit to the Hubbell House makes those immediately clear. Anthropologist Hall describes the trading post as a visual feast. So was the house. Magnificent Navajo rugs covered the floor of the spacious Hall, a combination living-dining room flanked on both sides by small, simple bedrooms where Hubbell put up his guests: teachers, missionaries, writers, artists, even notables like President Theodore Roosevelt. "Mr. Hubbell, you are a strenuous man," the president said after just a week with the trader, who had run *him* ragged for once.

Hubbell played host to hordes of visitors, including President Theodore Roosevelt, in this dining room.

For centuries, the Navajos had raised sheep and used their wool in weaving. The wool and the blankets were traditional trade items, so there was a base for economic development. But Hubbell urged the Navajos to improve the quality of their weaving to enhance its appeal to a broader market, and he encouraged them to weave blankets so large and heavy they could only be used as rugs. It was at his trading post that the "Ganado Red" originated and flourished. It's perhaps the best known of all Navajo rug styles, and how people think a Navajo rug should look. It was Hubbell, as much as any trader, who marketed the Navajo rug so skillfully that they are now prized collector's items around the world.

Hubbell knew the legendary Apache leaders Geronimo and Cochise well. The two most influential Navajo leaders of the day, Manuelito and Ganado Mucho (pidgin Navajo for "Many Cattle"), were good friends. "Many Horses," Ganado Mucho's son, is buried next to Hubbell in the family cemetery on Hubbell Hill, overlooking the trading post complex. So Hubbell's house was an ideal destination for writers like Hamlin Garland, who relished Hubbell's adventures and acquaintances as inspirations for his own

RIGHT: *The Main Hall of Hubbell's house today remains a treasure house of Native American art.*

OPPOSITE: *The Rug Room at the post is renowned. The famed Ganado red style that originated at the Hubbell Trading Post became the conventional idea of how a Navajo rug should look.*

stories. Young artists wanting to make their reputations in Indian country found their way to Hubbell's door. The foot-and-a-half-thick adobe walls of the Hall are covered with their work. Elbridge Ayer Burbank, the "Catlin of the West," or "Many Brushes," stayed a year. Two likenesses of the famous warrior Geronimo hang in the Hall: an oil painting of his head, and one of Burbank's famous red crayon profiles. Portraits of Chief Sitting Bull, Red Cloud, Chief Joseph, Manuelito, and many other epic Native American leaders by Burbank hang in Hubbell's remarkable portrait gallery. Visiting artists left Hubbell another unique collection: more than fifty miniatures of classic Navajo blanket

RIGHT: *The bedrooms in the house were simply furnished.*

OPPOSITE TOP: *The trading post itself was opened by Hubbell in 1876.*

OPPOSITE BOTTOM: *Don Lorenzo was photographed in the Main Hall around 1903. He served as a valuable link between the three cultures of the Southwest.*

patterns he had selected to serve as visual clues for the weavers at his post. William Robinson Leigh, "the Sagebrush Rembrandt," Maynard Dixon, and others also came for inspiration. Hubbell's collection includes their oils, watercolors, and pastels of Western landscapes, native ceremonies, hunters, weavers, and potters at work.

Every spare inch of wall and ceiling between ponderosa pine vigas is covered with objects Hubbell collected over the years not only from the Navajos, but from the Apaches, Utes, Hopis, Comanches, and other Plains Indians: baskets, wedding sashes, dance wands, saddlebags, prayer sticks, sheep bells, a beaded dance bag, moccasins, and a rare horsehair bridle set. One expert recently put it this way: "The Hall would blow museum people's minds!" It certainly impressed Hubbell's guests, who went away with a greater appreciation of Native American culture.

Critics then and now see flaws in what Hamlin Garland hated as an "intolerable monopoly." A trading post could hold its customers captive to a system of credit and scrip, metal currency that could be used only at their post. Traders paid by the pound for some rugs that had taken hundreds of hours to make, from the shearing of the wool to the carding, dyeing, and spinning. A 1973 time-ratio cost chart (sheep to rug) calculated that the weaver of the highest-quality rugs ended up getting $1.30 per hour for her labor. With the use of high-quality commercial yarn by 1988, the hourly price increased to $10.30.

Some have criticized the relationship between the Indian trader and his customers as paternalistic at best, feudal at its worst. Navajo workers looked back on Hubbell's wages as very low. But he did provide jobs and extend credit during hard times to many customers. He gave promising young Navajos important responsibilities in all aspects of his trading enterprise. Ironically, Hubbell died heavily indebted himself to Henry Chee Dodge, a wealthy Navajo and first president of the Navajo Nation, who had been able to prosper in both the Navajo and Anglo worlds. Hubbell's campaign for state and national elective office cost more than the always risky trading business and his extraordinary generosity could safely allow. His two sons, Lorenzo Junior and Roman, carried on the tradition well during their lifetimes, each in his own fashion, and the Hubbell Trading Post is one of the few posts, and probably the oldest, still in business today. It's now a National Historic Site.

"What changes . . . what times I've seen," Hubbell said just a few months before he died. "You who are younger will live to see even greater changes. We're only seeing the beginning." What was special in Hubbell's case, what has lasted, in the view of a former curator at the post, "is the goodwill that he engendered, a source of belief that it is possible for people of different races, languages and cultures to live and work together in harmony . . . if only they try."

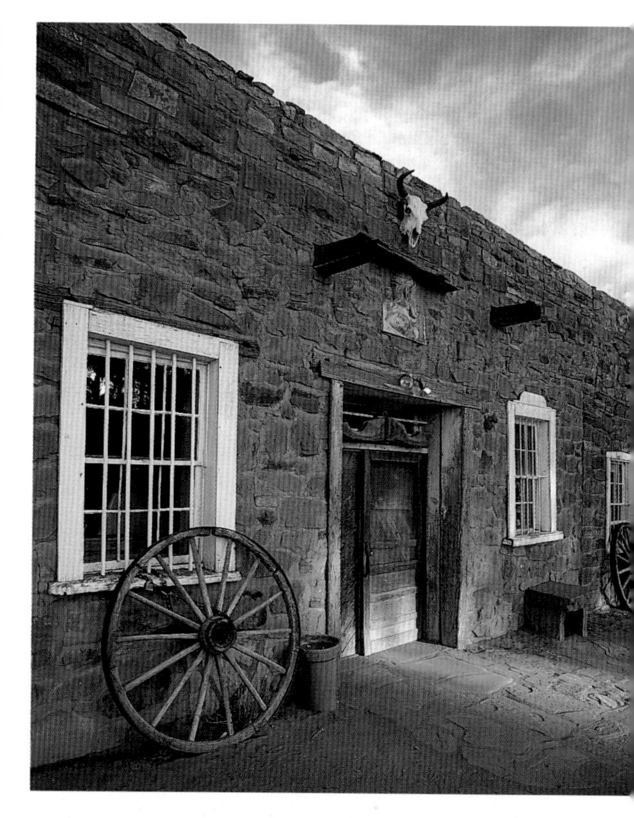

V

A WOMAN'S PLACE

Many readers will have heard of one woman portrayed in this chapter; in her day she was world-renowned. But few know that the place Eleanor Roosevelt called home for the last seventeen years of her epic life was a converted furniture factory. Fewer will know that the architect of Hearst Castle, built for one of the most notorious, extravagant men of the twentieth century, was Julia Morgan, a tiny woman in whom he put his trust for twenty-five years. Rebecca Nurse's Salem Village homestead serves as a symbol of courage and integrity in the face of death. Condemned as a witch, she refused to tell the lie that would have saved her life.

REBECCA NURSE HOMESTEAD

Danvers, Massachusetts

Rebecca Nurse had been sick in bed for "allmost a weak." She was so deaf she didn't hear her four visitors hurrying up the dark and narrow twisting staircase to the bedchamber above the Great Hall. Only one of her eight children was at the homestead that early spring day in 1692. Sarah dreaded letting the callers in, because she had a good idea why her mother's old friends had come. It was to warn her that she was in great danger. In the days before they made their way along the winding path that crossed the Nurses' fields, and up the knoll to her house, three village women—Sarah Good, Sarah Osburn, and Tituba, Reverend Samuel Parris's West Indian slave—had been examined in the village meeting house, then sent off in chains to prison in Boston to await trial "for Suspition of Witchcraft." Gentle Martha Cory had been committed "unto the Gaole in Salem" on March 21, for the same offense. Now Rebecca Nurse, of all people, was under suspicion.

OPPOSITE: *This narrow staircase twists up to Rebecca Nurse's bedchamber, the single room on the second floor above the Great Hall.*

PREVIOUS SPREAD: *The Rebecca Nurse Homestead in Danvers, Massachusetts, is a rare survivor of the First Period in American architecture (1620–1720).*

Goodwife Nurse had been as upset as anyone by what had been going on in Salem Village for weeks, "in perticuler" at Reverend Parris's parsonage. Rebecca and her husband, Francis, were among those villagers who didn't like Parris's policies and said so publicly. He was too embittered and demanding. But Rebecca was a compassionate woman, and a full covenant church member. As her friends sat on her stiff ladder-back chairs by the great fireplace in the bedchamber, she told them how sorry she was that Reverend Parris's nine-year-old daughter, Betty, was one of the girls who'd been bewitched. Their convulsions and screams were "Awfull to behold," Rebecca knew. "She pittied them with all her harte."

Prayers and fasting hadn't helped the girls at all, and what was worse, according to an eyewitness at Martha Cory's examination, "the Number of the Afflicted Persons" now included "Four Married Women . . . an Ancient Woman . . . and three Maids" as well as the "three Girls from 9 to 12 Years of Age" who'd started it all. Only one plausible explanation was left in the villagers' harsh Puritan belief system for what was "greviously tortoring" them. "The Evil Hand is on them," a village doctor, William Griggs, concluded. Reverend Parris's diagnosis was even more ominous: "The Devil hath been raised amongst us, & his Rage is vehement & terrible, & when he shall be silenc'd the Lord only knows." This was appalling news for Salem Villagers, for the Devil was as real to them as God himself. Those who agreed to do Satan's bidding were criminals, not only in God's eyes, but under the colony's law. Their crimes were punishable by death.

Rebecca Nurse, like Martha Cory, didn't fit the typical witch profile. True, she was old, frail, and female. There was talk her mother had been a witch. And hadn't Ben-

When Reverend Samuel Parris announced in March 1692 that "the Devil hath been raised amongst us," the witch hunt intensified.

jamin Holton died soon after she'd scolded him for letting his hogs get into her fields? These were incriminating signs. But in most respects Rebecca was the model Puritan wife and mother, not a crotchety, outspoken, middle-aged misfit like Sarah Good and Sarah Osburn. In a few years, her family would own 300 acres of precious farmland in Salem Village, where they'd moved in 1678. Their solid, wood-framed house, with its massive chimney and steeply pitched roof so like the architectural traditions of Rebecca's English birthplace, was an imposing sign of the Nurses' rising social and economic status, and their strength as a family unit. In the weeks to come, they would also show their enormous courage and deep love for the family matriarch.

The Great Hall of the Nurse homestead was the only room on the first floor in those days. It was 17 by 17 feet, spacious for the time. The fireplace was 7 feet wide, with two ovens inside. The walls were plastered white, and the ceiling whitewashed for light. As heads of the household, Rebecca and her husband would have slept here when they were in good health. The casement windows on both floors and in the garret were fitted with glass, not the oiled paper or sliding shutters poorer villagers could afford. By Salem Town standards, the Nurse home was small. Richer families in town might have two rooms on each floor, with bright colors on the joists and beams—yellow, perhaps, purple, or white. But the huge summer beams in the Nurse house, running on both floors from the front of the house to the back, and the window frames, did have ornamental carved touches.

When her visitors could finally bring themselves to tell Rebecca that she had been "spoken of allsoe" as a witch, she "sate still A whille being as it wear Amazed," her friends reported later. Then she said, "I am as Innocent as the child unborne but seurly . . . what sine hath god found out in me unrepented of that he should lay such an Affliction upon me In my old Age . . . ?"

On March 23, 1692, a warrant for Rebecca's arrest was issued, "for vehement Suspition, of haveing Committed Sundry acts of Witchcraft, and thereby haveing donne Much hurt and Injury to the Bodys of Ann putnam the wife of Thomas putnam of Salem Village, Anna putnam ye daufter of Said Thomas putnam, and Abigail Williams & c."

Rebecca was taken from her home early in the morning the next day. Her husband, their four sons, and four daughters fought valiantly for weeks to save her life. The children filed a petition for their "honored and dear mother [who led] a blameless life." Her husband got thirty-nine neighbors, at great risk to their own lives, to sign a petition: "To all whom it may concerne that . . . we never had Any: cause or grounds to suspect her of Any such thing as she is nowe Accused of. . . ." Rebecca filed two petitions on her own behalf. On June 28, 1692, she begged for a second physical examination by "Moast

The Great Hall was originally the only room on the first floor of the Nurse Homestead.

Grand wise and Skillful . . . Midwives," which would show that the Devil's mark "sum women" claimed to find on her body was "nothing . . . but what Might Arise from A natural Cause," i.e., having given birth to eight children. The jury at first found her not guilty, then changed its mind when "the honoured court had manifested their dissatisfaction of the verdict." The governor issued a reprieve, then revoked it, and left town. Nothing could save her; the hysteria was too intense.

Spectral evidence, what playwright Arthur Miller calls "that poisoned cloud of paranoid fantasy," was still admissible—only the accuser could see the accused person's invisible specter doing her harm. Reverend Parris's predecessor, Reverend Deodat Lawson, had not helped Rebecca's case when he visited her neighbor and accuser, Ann Put-

LEFT: *Seventy-one-year-old Rebecca was sick in bed when friends came to tell her she was under suspicion as a witch.*

BELOW: *Illustrator Howard Pyle depicted the Salem witch trials in this nineteenth-century engraving.*

OPPOSITE: *This casement window indicates the Nurse family's relatively prosperous standing in Salem Village.*

nam, to witness spectral evidence for himself. Lawson had come back to Salem Village to bolster the villagers' morale in their fight against Satan with a lengthy sermon. Goodwife Putnam began, he recounted later, "to Complain of, and as it were to Converse Personally with, Goodw[ife] N. saying Goodw[ife] N. Be gone! Be gone! Be gone! Are you not ashamed . . . to afflict a poor Creature so? . . . be gone, do not torment me. . . ." Rebecca, of course, was home in bed at the time. She was examined the next day, but refused to confess to any of the charges, as Tituba had, knowing that if she went along with the charge of witchcraft, her chances of being hanged, at least right away, were greatly reduced. "Would you have me bely myself?" Rebecca asked.

Goodwife Nurse never saw her home again. On July 19, 1692, she was hanged by the neck, and slowly strangled to death. Soon after, Rebecca's family brought her body back home secretly at night by boat up the brook that ran below the house. No one is sure exactly where she lies in the family graveyard close by the homestead.

After eight months of terror, the witchcraft hysteria died down. Rebecca would come to represent the beginning and the end of the deadliest witch hunt in American history, but not the last one. Once a woman of Rebecca's standing was targeted, men and women further up the Puritan hierarchy became fair game. Priscilla and John Alden's grandson was one; the richest man in Salem, another. Talk of the governor's

A monument to Rebecca Nurse stands in the family graveyard near the house. No one is sure exactly where her body is. John Greenleaf Whittier wrote the epitaph:

Oh Christian martyr
who for truth could die
when all about thee
owned the hideous lie!
The world redeemed
from superstition's sway
Is breathing freer
for thy sake today

wife being a witch was the last straw. When the court threw out the validity of spectral evidence, the witch hunt was over. But by the time the accusers were discredited, spectral evidence renounced, and juries made more broadly representative, 150 people had been accused, examined, and imprisoned in a number of Massachusetts towns north and west of Boston.

Those who had confessed, like Tituba, were never tried, and so survived; some had escaped to other colonies; the threat of a huge lawsuit by a Boston man for defamation squelched another. No one was listening to the accusers by the end. But nineteen innocent people, six men and thirteen women, including seventy-one-year-old Rebecca and her sister, Mary Esty, fifty-eight, had been executed; an eighty-one-year-old man, Martha Cory's husband, Giles, had been tortured to death; and five had died in prison. Tavern-keeper John Procter had made the fatal mistake of saying a good whipping would cure what ailed the girls. Poor homeless Sarah Good was "turned off" the ladder the same day as Rebecca. Her infant daughter had died in jail some time before. So had Sarah Osburn. Martha Cory survived harsh prison conditions until September 22, 1692.

Mary Esty fought as courageously as her sister, Rebecca, but she would hang from the gallows with Martha Cory. Dorcas Good might just as well have died. She was only five when "cried out upon" as a witch. Years later, her father said, "Being chain'd in the dungeon [she] was so hardly used and terrifyed that she hath ever since been very chargeable having little or no reason to govern herself."

The people of Salem Village, now the town of Danvers, where it all began, have preserved the Nurse Homestead and the twenty-seven acres surrounding it, as a haunting and stark symbol of that cruel and complex time. They have no simple explanation of what went wrong. Was it jealousy that started it all? Revenge? Greed? Sexual repression? Clinical hysteria? Food poisoning? Or was it the extraordinary tensions of the period: smallpox, the threat of Indian attack, new uncertainties about who owned what pieces of land, bitter divisions over how to meet Reverend Parris's demands, or frustration over attempts to free the village from the dominance of nearby Salem Town? One descendant of Rebecca Nurse says, "You come away feeling that no explanation is satisfactory. It is too much to comprehend."

There is no doubt the institutions of the day failed to protect the innocent, that some who should have been most charitable were the most vindictive. When Reverend Parris at long last prayed that "all might be covered with the mantle of love and we may . . . forgive each other heartily," one of Rebecca's sons-in-law replied, "If half so much had been said formerly it would never have come to this."

The Danvers Alarm List Company, a re-created eighteenth-century militia unit, now owns the Nurse Homestead. This small, modest group of townspeople has chosen deliberately to keep its commemoration of 1692 simple. They've had a constant fight against the commercialization and trivialization of witchcraft that now bedevils nearby Salem. There's not a broomstick in sight in Danvers, except perhaps the one in the Nurses' lean-to kitchen. To them, the Rebecca Nurse Homestead serves best as a poignant warning that "each generation must confront its share of scares and witch hunts with integrity, clear vision and bravery."

JULIA MORGAN'S HEARST CASTLE

SAN SIMEON, CALIFORNIA

ON AUGUST 12, 1926, JULIA MORGAN RECEIVED AN EXTRAORDINARY letter from a client for whom she was building a castle. "Dear Miss Morgan," it began. "How about a maze in connection with the zoo. I think getting lost in the maze and coming unexpectedly upon lions, tigers, pumas, panthers, wild cats, monkeys, mackaws and cockatoos, etc. etc. would be a thrill even for the most blasé. . . ." The writer was the notorious newspaper publisher, film magnate, and art collector, William Randolph Hearst. Julia Morgan was his architect. For a quarter century, this unlikely pair worked together to create his hilltop extravaganza—not one but four houses and more—at San Simeon on the Pacific coast 200 miles south of San Francisco.

This particular Hearst idea—a zoo—wild as it seems, didn't faze Miss Morgan. Very little did. She had long since proved to be one of the most versatile architects in the United States, with over 350 commissions already completed, several for Hearst's formidable mother, Phoebe Apperson Hearst, and for himself so far, a building in Los Angeles for his newspaper, the *Examiner*. Morgan's success hadn't come easily. There wasn't one school of architecture in the West in 1890, so she got a degree in engineering at the University of California at Berkeley. The architecture division of the École des Beaux Arts in Paris had never admitted women in the two centuries since its founding in 1691. In 1902, Morgan would be the first to receive a certification in architecture. She had failed the entrance exam twice in two years, one outraged mentor insisted, simply because the jury "did not want to encourage young girls." And before she set up her own practice in San Francisco in 1904, she had worked for an architect in Berkeley who bragged she was "the best and most talented designer, whom I have to pay almost nothing, as it is a woman."

ABOVE: *In 1902, Julia Morgan became the first woman to receive a certificate in architecture from l'École des Beaux Arts. She designed 700 buildings during her prodigious career.*

OPPOSITE: *Behind the limestone facade of the Main Building is a reinforced concrete, earthquake-resistant building, with at least 115 rooms.*

Many people feared William Randolph Hearst. He was rich and ruthless. "He could make or break you and if you didn't do as you were told, why you could be just [sent] right down off that hill." "That hill" was a remote, wild, barren place, 1,600 feet above sea level, a high point on the great ranch Hearst's father, George, had acquired with his mining millions. The Hearst family had gone camping on the hill for years in a kind of tent city. But Willie, George and Phoebe's only child, was finally in control of their millions by 1919, and wanted "something that would be more comfortable," he told Miss Morgan in her private office one April afternoon in 1919.

"The experts told Pop it couldn't be done. They told the old man to forget it," his son recalled years later. Neither Hearst nor the site intimidated Julia Morgan. She surveyed that in August 1919, and incredible as it seems, just five weeks after her initial site

Julia Morgan employed skilled craftsmen in wood, stone, iron, tile, and glass to incorporate Hearst's art collection into the construction, and to create new details.

visit, she was ready to begin construction. The odds against her were enormous. The nearest railroad was fifty-six miles away. The wharf on San Simeon Bay, built by George Hearst in 1878, could not accommodate ships large enough for the construction equipment Morgan needed to bring in. There were no roads up to the hilltop, just cattle trails. There were few workmen in the area, and fewer raw materials. There was no place to store the thousands of crates containing part of Hearst's enormous art collection "brought from the ends of the earth and from prehistoric down to late Empire in period," Morgan observed. But just two years later, she wrote to one of Hearst's antique buyers in Europe, "We are building for him a sort of village on a mountain top overlooking the sea and ranges of mountains . . . and housing incidentally his collection as well as his family." Morgan was on the site almost every weekend for the next eighteen years. Her office in San Francisco was a complicated overnight trip away, but she was always back there on Monday mornings, working on other commissions which topped seven hundred by the time she retired in 1951, at age seventy-nine.

Hearst always wanted more. "The big house is a whale," he wrote Miss Morgan in March 1924. "If I'd known it would be so big, I would have made the little buildings bigger." Morgan always managed to give Hearst more. She designed two magnificent "plunges": the outdoor Neptune Pool took twelve years to perfect, the indoor Roman Pool seven years. She created a mile-long pergola ("the longest in captivity," she mused), and a zoo for what may have been the largest private menagerie in the country. She built kennels, hothouses, planted stands of thousands of pine trees, and glorious gardens. Counts still vary of the grand hilltop total: from 127 to 165 rooms in the Casa Grande and the three "little buildings," or guest cottages; as many as 58 bedrooms, 61 baths, 18 or 19 sitting rooms, and 2 libraries, all "furnished with antiques and architectural elements–many original, many done on the spot" by the stonecutters, woodcarvers, and other skilled craftsmen Morgan engaged, like the medieval master builders she admired.

Hearst and his longtime love, the popular actress Marion Davies, liked to entertain world celebrities at San Simeon.

George Bernard Shaw was one of the famous guests who received a coveted invitation to spend a weekend at Hearst Castle with Hearst and a score or more of other guests. "This is the way God probably would have done it if he'd had the money," Shaw is reported to have quipped. Hearst's longtime companion, Marion Davies, a clever, rich, and popular film star, wasn't inclined to be impressed by anybody or anything. Hearst turned pale at the sight of her friends dancing on top of a priceless medieval chest. She dismissed Wyntoon, a Bavarian village complex Julia Morgan began for Hearst in 1924 in northern California, as "Spittoon." But even she admitted in her memoir, *The Times We Had,* that San Simeon was awesome and Versailles couldn't compare.

Julia Morgan spent little time with Hearst's glitterati. She was too busy "running the job," as she called it. Sara Boutelle, Morgan's biographer, lists

> hiring, firing, and settling disputes; arranging lodging, food and working quarters for the laborers; making trips to interview specialists such as a cheese maker, a chicken man, gardeners, and housekeepers; procuring special plants and materials; creating various crafts centers on site; arranging transportation by ship, rail, and truck to the remote hilltop; building warehouses and cataloging objects to be incorporated in the project; checking on thousands of details; and satisfying the whims of artists and the client.

San Simeon was a colossal, unique achievement for Julia Morgan, all five feet and 100 pounds of her. The grand mansions architect Richard Morris Hunt built for all the Vanderbilts combined didn't come close to what she handled at San Simeon alone. Morgan didn't make much on the project, however–about $71,000, biographer Boutelle estimates. That's a little under $3,000 per year for the twenty-five years she devoted to it.

RIGHT: *Winston Churchill, the Duke and Duchess of Windsor, Charles Lindbergh, Greta Garbo, George Bernard Shaw, Charlie Chaplin, and many other rich and famous visitors may have used this guest library.*

BELOW: *Morgan and Hearst worked closely together for a quarter century on this gigantic project, but never finished it. She designed other buildings and complexes for Hearst and his mother, Phoebe.*

Even though she was working on other commissions at the same time: private homes, apartment houses, hospitals, libraries, orphanages, museums, and churches, she had very little money in her sad, reclusive final years.

Morgan was Hearst's opposite in many ways. He was a huge physical presence. She was "a neat bantam hen among peacocks." To her, "money was a small part of living." To Hearst, it meant everything, not for itself but for what he could buy and build with it. For years, he was said to be the world's biggest spender—an average of a million dollars a year on his art collection alone—and was $126 million in debt by 1937. Julia Morgan was austere, ascetic, dressing in French silk blouses and tailored suits. He had gargantuan appetites, flamboyant tastes. Even at the age of ten, traveling in Europe with his mother, he said he'd like to live at Windsor Castle, then asked her to buy the Louvre! Morgan preferred anonymity; she refused to enter competitions, write articles, serve on committees, and she saw no reason not to destroy most of her files, blueprints, and drawings when she closed her office. Hearst was shy personally, but he was a megalomaniac, a genius at publicizing and promoting his enterprises, although the comment he is supposed to have made to his war correspondent in Cuba ("You furnish the pictures, I'll furnish the war") is now discredited.

The Main Building (La Casa Grande) overlooks a complex that included three guest cottages, two mammoth swimming pools, a mile-long pergola, hothouses, and kennels.

But they had much in common beside their San Francisco origins. "Mr. Hearst and I are fellow architects," she said. "He loves architecturing. He supplies vision, critical judgment. I give technical knowledge and building experience." They were both prodigious workers. To her, an eighteen-hour day was nothing. She had the constitution of an ox and climbed scaffolding like a goat, her nephew said. She had no private life, and could go for days on little sleep, candy bars, and coffee. At San Simeon, Hearst worked most of the night in his third-floor Gothic Suite, turning the pages of his twenty-six newspapers with his feet, calling his staff at all hours from any of the eighty telephones on the grounds. Hearst and Morgan were both fearless. Once Morgan went down a drainpipe at Wyntoon and came out two miles away. "Everyone was looking for her above ground," a colleague said. They could talk for hours about San Simeon, and bought two copies of books so they might confer easily when they were apart. They both loved California, its natural beauty, and architectural heritage. Both were passionate about the details of their profession. Hearst was completely familiar with the minutest aspects of each of his many enterprises. She designed and engineered everything at San Simeon, from fireproof and earthquake-resistant structures down to the iron grilles guarding his rare books and formulas for demothing his precious tapestries. One Morgan associate said, "They were both long distance dreamers, looking way, way ahead" to the prospect of San Simeon's becoming a museum of the best that money could buy. Hearst had no qualms about changing his mind, even if it cost thousands of dollars to tear down a tower already built, or move an ancient oak tree a few feet so his guests would not have to duck. This wasn't easy for Morgan, but she never showed her anger, to him, at least. She admitted her own "changeableness of mind," and knew how to handle Hearst's suggestions, however preposterous.

> I like your idea for the combination indoor pool and orchid greenhouse. It should be very tropical and exotic. There could be a plate glass partition in the pool, and the alligators sharks, etc. could disport themselves on one side of it, and visitors could unsuspectingly dive toward it.

Julia Morgan understood what Hearst was groping for, her nephew explained, and it was "probably a windfall in a sense, because she had the training for what he had in mind, and the imagination and the ability."

Whether she was designing for Hearst, a college professor, or for one of her top workmen at San Simeon, Julia Morgan tried to satisfy their needs and desires. She insisted on private dining rooms and kitchenettes in the new YWCA in San Francisco so the young women could entertain friends at meals. "But these are minimum wage girls,

why spoil them?" one critic asked. "That's just the reason," Morgan replied. She would sit on the floor with a client's children before drawing up plans, and made a point of putting something special in each house just for the children–a hiding place, or secret steps. She designed a playhouse for the daughter of the taxi driver who drove her to Hearst Castle over the years, and she shared profits, when there were any, with her staff. Hearst paid top salaries to his workers, and enjoyed touring Europe for weeks at a time with an all-expenses-paid entourage, buying clothes for everyone if the weather changed. Once, so the story goes, he replaced a goose he'd run over on a country road with a new one, delivered to the owner in a new automobile! He'd give his prized dachshunds away to guests for the asking.

This magnificent indoor "Roman" pool took seven years to finish. The tiles were made of 22-karat hammered gold and Venetian glass.

During his career, Hearst was called "a burning disgrace to the craft." But to Lincoln Steffens, the muckraker, "the publisher had far more depth and substance than he was given credit for." That was the side Julia Morgan chose to see. Toward her, he was charming and respectful, not the odious person biographer Boutelle expected to find before reading their correspondence. He could reveal his weaknesses to Morgan. "Heartily approve those steps . . . I certainly want that *saisissant* effect. I don't know what it is, but I think we ought to have at least one such on the premises."

ABOVE: *Just about everything at William Randolph Hearst's castle on "The Enchanted Hill" at San Simeon, California, was designed by the architect Julia Morgan—even the world's largest private zoo.*

LEFT: *Hearst worked long hours in the library of his private Gothic Suite on the third floor of the main building. The only portrait for which he ever sat is at the rear of the room.*

Hearst's regard for Miss Morgan was clear and constant. Her telephone calls were the only ones he accepted while he was out in the gardens at San Simeon. When apart, they were in touch sometimes two or three times a day. She sat next to him at meals in the enormous Casa Grande refectory. "He just lived for plans," Marion Davies said. "Anytime you wanted to find him, he would be in the architect's office," a tiny shack behind the Casa Grande, and it's still there. "You are the best judge of that," he told Miss Morgan many times. "I will leave the matter in your hands." And she was willing to pay the price, her nephew said, "to be able to build things on a scale that in private enterprise would be virtually impossible."

Together they created a romantic, surprisingly delicate "Enchanted Hill." In the few photos that exist of Hearst and Morgan together at Hearst Castle, she is identified as his secretary, if at all. One critic concludes that "this great Californian deserves in American architecture at least as high a place as Mary Cassatt in American painting, or Edith Wharton in American letters." That is the least that can be said of this intrepid genius.

ELEANOR ROOSEVELT'S VAL-KILL COTTAGE

Hyde Park, New York

Val-Kill Cottage, a converted furniture factory, was an unlikely home for the most famous woman in the world, but she loved it.

Eleanor Roosevelt had no qualms admitting that her house was "somewhat odd." It was, after all, a converted furniture factory. By 1936, the business she'd run there for close to a decade had failed, like so many others during the Great Depression. She didn't take long to decide to turn the "perfectly good" structure into a "fairly comfortable" home. It was still a bizarre, rambling, two-story stucco warren of nooks and crannies, small rooms, and narrow hallways. But to her, it was perfect. She was fifty-two by then, and at last she had a home of her own.

When Mrs. Roosevelt moved into Val-Kill Cottage in 1937, she had been First Lady for four years. For the next eight years, the cottage was her refuge, not only from the White House, which, unlike Mary Todd Lincoln, she had never coveted, but from the persistent miseries of living at her mother-in-law's house, Springwood, just two miles away through the woods. After President Franklin Roosevelt died soon after the start of his fourth term in the spring of 1945, Val-Kill Cottage became her permanent residence for the last seventeen years of her epic life.

"The peace of it is divine," she had written Franklin about the site years before. "I could be happy in the house alone," she told her daughter, Anna. Mrs. Roosevelt's perspective on peace and privacy was relative. One day at the White House, she had to cope with 30 guests for breakfast at 6:30 A.M., 35 for lunch, 400 for tea at 4, 500 for tea at 5, 80 for dinner, 300 for an evening musicale, champagne punch at 11 P.M., and a snack for the musicians after that.

Val-Kill sometimes seemed just as busy. "She loved having company, just loved it," said Marguerite Entrup, her last cook, who had learned to be ready on short notice for breakfast for "only 14 tomorrow, Marge," lunch for 40, or picnics for 150 (and the rolls had to be buttered just as carefully for the delinquent boys from Wiltwyck School as they were for visiting royalty). In her autobiography, Mrs. Roosevelt wrote, "My mother-in-law once remarked that I liked to 'keep a hotel.' I probably still do when I am at Hyde Park." Maureen Corr, her last secretary, saw it differently. "She could not bear to be alone."

"Possessions seemed of little importance, and they have grown less important with the years," Mrs. Roosevelt said as she began life on her own in 1945. She did treasure her Irish linen sheets and tablecloths, but the dining room chairs at Val-Kill Cottage didn't always match, the lamp shades were usually crooked, the dime store glasses looked incongruous on her dining room table next to the heirloom silver candelabra. That did not matter to her–or to those who admired her as the most dedicated, outspo-

Eleanor Roosevelt was twenty when she married her distant cousin Franklin in 1905. She was given away by her Uncle Ted, then president of the United States. The marriage endured, a difficult but respectful and powerful partnership.

ken humanitarian of her time. The house was comfortable, welcoming. "You felt at peace there," says a close friend, who remembers the bouquets of handpicked flowers Mrs. R. liked to leave in each guest's room. The lack of pretense at Val-Kill didn't surprise those who had seen her carrying her own bags, or standing in line at the U.N. staff cafeteria. The emperor of Ethiopia felt enough at ease at Val-Kill to take off his shoes while he and his retinue watched an earlier television appearance. Prime Minister Nehru of India sat on the living room floor in his immaculate white trousers to talk to young guests, whose views and future always concerned Mrs. Roosevelt deeply.

Some people detested her as much as they did her husband—"the most dangerous woman in America," they believed. But by 1940, Mrs. Roosevelt had become more popular in the Gallup polls than her husband. A bomb threat where she was to speak didn't bother one admirer. If she had to be blown up, the woman said, there was nobody she'd rather be blown up with than Mrs. Roosevelt.

Val-Kill was a must stop for royalty, presidential hopefuls, cold warriors, contrite cardinals, movie stars, labor leaders, and the countless individuals she had befriended on her global quests to learn and understand. "Driving thro' cheering thousands doesn't make me feel I know much about the people & it's the people who really interest me," she said. "She took you as she saw you," said Marge Entrup. Her great-granddaughter, Nina, remembers her saying, "There isn't anybody you can't learn something from."

Mrs. Roosevelt's belief in the need for roots seems paradoxical. "I'm always given the reputation of being constantly on the move," she said, and her world travels were constant grist for cartoonists and critics. She could turn up anywhere—at the bottom of a mine shaft, at a migrant worker's hovel, a locked ward for soldiers who'd gone mad in World War I, even an embattled island in the South Pacific in World War II. People joked that Admiral Byrd set two places for dinner every night at his Antarctic hut just in case she dropped by. Her Secret Service code name was Rover.

Nevertheless, she was convinced that "all of us need deep roots. We need to feel there is one place to which we can go back, where we shall always be able to work with people whom we know as our close friends and associates, where we feel that we have done something in the way of shaping a community, of counting in making the public opinion of the community." The Hudson River Valley was that place, as precious to her as it was to her husband.

Roots had been lacking during a childhood she described as "very miserable." "Where is baby's home now?" she asked an aunt when she was only two and a half. When she was eighteen, in tears she told another, "I have no real home." Her distraught, cold, beautiful mother, Anna, and her adored, alcoholic, exiled father, Elliot Roosevelt, both died before she was ten. Except for the three years she spent at school in England ("the happiest years of my life"), she lived with eccentric relatives in a succession of houses as grim as they were grand. When she was just twenty, she accepted the proposal of her dashing fifth cousin, Franklin, only to find that every house they would ever live in—except the New York State Governor's Mansion and the White House—was under the domination of Franklin's mother, Sara. Once Val-Kill Cottage was hers, she stayed there as much as possible, returning to Springwood at night for the sake of appearances only if her husband was there. He, on the other hand, loved to come over to Val-Kill Cottage for a swim, a picnic, or a rare supper alone with her.

This classic image of Mrs. Roosevelt carrying her own bag is a symbol of her modesty and lack of pretense.

Eleanor Roosevelt was disciplined, organized, ascetic. "Perhaps we need again a little of that stern stuff our ancestors were made of," she believed. Her day at Val-Kill began at 7 A.M. after a few hours' rest on her precious sleeping porch—until snow fell. "I like the still nights . . . with only the stars to look at just because it gives me a feeling of taking in. . . ." After a cold shower, calisthenics, and a walk with the world-famous Scottie, Fala, or his son, Tamas, no matter what the weather, she was ready to join family and guests for breakfast, promptly at 8:30. Illness was the only excuse for not being there and on time. Lunch was served at 1, dinner at 7 or 7:30 sharp, and she always changed for dinner. Eleanor detested alcohol, or loss of self-control, which was understandable: her father, her younger brother, and her uncle were alcoholics who had tormented those around them for years and finally destroyed themselves. Cocktails were served before dinner, but ten or fifteen minutes was enough, in her opinion. Meals were simple. Mrs. R. was far more interested in listening to the answers guests gave to her questions "that made people think," remembers Nina, who lived in nearby Stone Cottage, the first house on Val-Kill grounds, built by Franklin Roosevelt.

Mrs. Roosevelt's day ended hours after she had said good night to her guests. She would rise around 10 P.M. from her chair beside the living room fireplace and go back to work. Ever since World War I, which she regarded as her emancipation and education, she had been a prodigious worker. "Dear God, please make Eleanor a little tired," was said to have been her husband's nightly prayer. But because she didn't give up, others couldn't, remembers Maureen Corr. Mrs. Roosevelt's output was enormous: seven books, a syndicated column, "My Day," for over twenty-five years, countless letters in longhand to her family and friends (2,336 to one friend alone). She was always in great demand as a speaker. More than once she was called on as the only person respected and powerful enough to defuse a dangerous situation—at the Democratic convention of 1940, at the U.N. during the Cold War. Edward R. Murrow called her 1956 speech the greatest convention speech he had ever heard.

She preferred to use her sleeping porch until the snow fell.

Mrs. Roosevelt collected friends, and just part of her living room reveals her enormous collection of photographs and mementos.

The large desk, the dresser, several chests of drawers, and side tables in her secretary's Val-Kill office were handsome examples of the furniture that had been made by local youths at the Val-Kill factory: sturdy colonial reproductions, stained and rubbed to a beautiful soft patina, as were the pine walls in her living room, dining room, and halls. The furniture factory was just one example of Eleanor Roosevelt's compassion and empathy. She always tried to put herself in other people's shoes. "Human beings are poor things. Think how much discipline we need ourselves." Her devoted secretary, Malvina "Tommy" Thompson, always worried that Mrs. Roosevelt would be taken advantage of, but Mrs. Roosevelt didn't care. During the Depression, she had picked up a starving young hitchhiker, sent him to her New York City apartment for a meal, found him work with the Civilian Conservation Corps, invited "her tramp" to Val-Kill, and years later was still sending his daughter (her namesake and goddaughter) $10 on each birthday. When a boy made a nameplate for her Val-Kill desk but misspelled her first name ELANOR, she accepted it graciously rather than point out his mistake, and it's still there. A seedy woman selling dead chickens came to the cottage door one day. "You never sent anyone away," Marge Entrup recalled. "Mrs. R. always had to be informed." She went to the door and bought them all. "This is the type she was, not afraid." Implausible as it seems, Eleanor Roosevelt had to overcome many fears in her life—of water, of speaking her mind, of parenting. Always shy and insecure, she knew, looking back on her life, that "we do not have to become heroes overnight. Just a step at a time, meeting each thing that comes up, seeing it is not as dreadful as it appeared, discovering we have the strength to stare it down." She had finally become, she said, "a tough old bird."

Eleanor took few things from Springwood before the house was turned over to the nation. Her husband's vast collections went to his new presidential library. Friends

Mrs. Roosevelt's office at Val-Kill Cottage was as unpretentious as the rest of it. The emperor of Ethiopia took off his shoes here to watch television with his retinue.

Mrs. Roosevelt's simple bedroom contained eight pictures of her husband.

Eleanor and Franklin lost one child in 1909, a second son, FDR Jr. In 1919, when Roosevelt was secretary of the navy, they posed with their five surviving children, a girl and four boys.

were her passion. Photographs and mementos of the people she cherished are everywhere at Val-Kill Cottage: hundreds of framed items and knickknacks on the walls, on tabletops, windowsills, over doors, on mantels; pictures of her four sons, their many wives and children, daughter Anna and her family, "Uncle Ted," the twenty-sixth president, even her first dancing partner. Photographs of Louis Howe, her husband's devoted political adviser, are on the piano and above it, in the living room. She and Louis were "the two homeliest people in New York politics," she commented. Eleanor grew to love his "rather extraordinary eyes and fine mind." Together, for seven years they kept FDR's political name alive while he struggled to come to terms with polio. With Louis's guidance, Eleanor became one of the shrewdest politicians in the country. "He always wanted to make me President when FDR was thro," she told a close friend, "& insisted he could do it. You see he was interested in his power to create personages more than in a person, tho' I think he probably cared more for me as a person as much as he cared for anyone & more than anyone else ever has!"

Over the years, she thought deeply about love and happiness. "What a nuisance hearts are," she said, "and yet without them life would hardly be worthwhile." Her own heart had been broken–by her father's death, her third baby's death, and then by the discovery of her husband's infidelity. But their marriage endured, and their political partnership became a powerful national force. When he died, Eleanor wrote, "My husband and I had come through the years with an acceptance of each other's faults and foibles, a deep understanding, warm affection and agreement on essential values. We depended on each other." She kept eight pictures of FDR in her Val-Kill bedroom alone.

She once begged a dear young friend "not to accept a half loaf of love." She had done that, but some who knew her husband well concluded that he was incapable of the emotional intimacy she craved. Roosevelt's admiration for his wife was deep. He told their son James she was the most remarkable woman he had ever known, the smartest, the most intuitive, and most interesting. Young and naive as he was when he proposed to Eleanor in 1904, he had sensed how special she was, just as had her father, her aunts, and, above all, her headmistress at school in England. Years later at Val-Kill, Mrs. Roosevelt told a friend, "I remember allowing my cousin Franklin to come down to meet me" (at a Lower East Side settlement house). "I wanted him to see *how people lived* . . . and it worked. He saw how people lived and he *never* forgot." Different as Eleanor and Franklin were temperamentally, they shared that compassion. "Through the whole of Franklin's career," she wrote, "there never was any deviation from his original objective—to help make life better for the average man, woman and child."

She never wavered herself, sometimes badgering her husband mercilessly when he gave way to political realities. He saw what could be done, someone said; she saw what should be done. To her son Elliot, she was an "undaunted idealist," and she fought to the end to close the gap she found between the democratic ideal and the American reality. "We have to prove . . . that democracy really brings about happier and better conditions for the people as a whole," she insisted. As chairman of the United Nations Commission on Human Rights, Eleanor Roosevelt worked brilliantly and successfully for passage of a Universal Declaration of Human Rights in 1948. World leaders tried for years to secure her the Nobel Peace Prize. "If she didn't earn it," said Harry Truman, "then no one else has." Later, President John F. Kennedy, who had come to Val-Kill urgently seeking her endorsement for his presidential campaign in 1960, wrote the Nobel Committee, "She was a living symbol of world understanding and peace, and her untiring efforts had become a vital part of the historic fabric of this century."

Typically, Eleanor Roosevelt put it more simply: "Where after all do universal human rights begin? In small places close to home—so close and so small that they cannot be seen on any maps of the world." Val-Kill Cottage is a perfect example of what she meant.

At nearby Stone Cottage, where she first learned to love the site, the Eleanor Roosevelt Center at Val-Kill (ERVK) now works to carry on her humanitarian legacy, and selfless example of leadership. The two buildings and surrounding 179 acres of grass, woods and water, are now a National Historic Site, the only one dedicated to a First Lady, not only in the United States, but in the world.

VI

CASTLES IN THE SAND

In their own spectacular ways, each "house" in this chapter tells of the fulfillment of dreams—very different dreams, yet all variations on that same persistent, thriving phenomenon, "the American Dream."

But these three houses—a Lower East Side tenement, a royal palace, and a 250-room château built for a bachelor—are also monuments to the tenuous, constantly shifting nature of power and wealth in America, how one dream dies, only to be replaced by another.

Behind the cast-iron columns and massive portals of ʻIoloni Palace, home of the last two monarchs of the Hawaiian kingdom, was a building equipped with the latest in plumbing, lighting, and communications technology (see ʻIolani, page 186).

BILTMORE ESTATE

ASHEVILLE, NORTH CAROLINA

"NOW I HAVE BROUGHT YOU HERE TO TELL ME IF I HAVE BEEN doing anything very foolish."

A fortune and great reputations were riding on the answer to that simple request. But George Washington Vanderbilt could not have known that, one fateful summer day in 1888, on a remote mountaintop in North Carolina. George was the youngest child of the richest man in America, and grandson of the great "Commodore." He had just finished buying thousands of acres of land in the western corner of the state. The family friend whose judgment he sought was the venerable landscape architect Frederick Law Olmsted, whose answer might have discouraged lesser young men: "The woods are miserable," he said, "the hillsides are so eroded they are unsuitable for anything that can properly be called park scenery. The soil seems to be generally poor."

ABOVE: *Thirty-three-year-old George Washington Vanderbilt was still a bachelor when his 250-room château opened in 1895, to his family's astonishment (portrait by John Singer Sargent, 1895).*

OPPOSITE: *Vanderbilt's French Renaissance-style château, Biltmore, epitomized the Gilded Age and remains the largest private residence built in the United States.*

But George, in his mid-twenties, was now worth millions. He could afford a big dream, and Olmsted would help him shape one. An elusive, enigmatic figure even now, George wasn't interested in the family business or its conventional social aspirations. That may explain why his father had left him only $10 million ($5 million of it in trust). Olmsted found George "delicate, refined, and bookish . . . with considerable humor," a lover of the arts and nature who really did want to do something worthwhile with his money. So what Olmsted went on to say fell on trusting and intrigued ears. "Such land in Europe," Olmsted told George, "would be made a forest; partly, if it belonged to a gentleman of large means, as a preserve for game, mainly with a view to crops of timber. That would be a suitable and dignified business for you to engage in; it would, in the long run, be probably a fair investment of capital, and it would be of great value to the country to have a thoroughly well organized and systematically conducted attempt in forestry made on a large scale. . . ."

Olmsted was at the end of a prodigious career. New York City's Central Park was just one of his many triumphs. What he was proposing to George was in fact the fulfillment of a vision he'd been fighting for all his life: thoughtful stewardship of the land and its resources. Olmsted had traveled on horseback over much of America, and had seen devastated landscapes like George's everywhere. This had been the American way since colonial times. From the Atlantic inland, it had been easy to mine the riches at hand, then move on. Here was Olmsted's last chance to put his theories of conservation into practice.

All George had in mind when he bought all that land was the wilderness view from the front porch of the modest hideaway he planned to build. So he told Olmsted

ABOVE AND OPPOSITE *Before: The construction of Biltmore was a mammoth six-year project: from 1892 to 1895, as many as 1,000 men worked six days a week, ten hours a day. After: The completed front entrance to the French Renaissance-style château features a staircase inspired by the King Francis I wing at the château at Blois, France.*

BELOW: *Vanderbilt posed on site with his distinguished collaborators: landscape architect Frederick Law Olmsted (with beard, sitting) and Richard Morris Hunt, the architect (leaning against the tree).*

he could "make a park out of it, I suppose." With that diffident remark, he'd hit at the heart of Olmsted's conviction that civilization depended on man's exposure to nature's beauty. Olmsted was especially concerned about the hordes who kept crowding into the nation's growing industrialized, ugly cities. Parks would help them keep their humanity. That was hardly the problem here, so Olmsted went on, "Make a small park into which to look from your house; make a small pleasure ground and garden; farm your river bottom chiefly to keep and fatten livestock with a view to manure; and make the rest a forest, improving the existing woods and planting the old fields." George was convinced. A year later, Olmsted was ready with a plan for George's mountains, hills, rivers, forests, and open land, in as unlikely a place to bring a dream alive as anywhere.

In the meantime, George had been traveling through Europe with the second giant he'd chosen to complete the dream Olmsted had inspired. Richard Morris Hunt was a charmer, the dean of American architects in the Gilded Age, anathema to only a few contemporaries. George knew Hunt well, too. They'd already worked together on George's first gesture as a young philanthropist, the design of a library that George then gave—land, building, and books—to New York City. George was almost like a son to Hunt, who felt he and George worked in perfect harmony. Here again, George started out against the family grain. He had something rather simple in mind for his North Carolina retreat, definitely not the opulent Fifth Avenue mansions and Newport cottages Hunt had come up with for others in the family. But Hunt, like Olmsted, was at the pinnacle of his career, and here was *his* chance to create a final masterpiece.

Hunt's earliest designs were conventional, but some 2,000 architectural drawings later, he had conceived, and George had approved, the construction of a French Renaissance château with a front facade 780 feet long, four acres of floor space, and at least 250 rooms, including 34 bedrooms, 43 bathrooms, 65 fireplaces, 3 kitchens, and an indoor swimming pool. And George was still a bachelor! He would not marry Edith Dresser until 1898, three years after Biltmore opened.

As the house rose on Lone Pine Mountain, where George had first asked Olmsted what to do, Hunt wrote to his wife, Catherine, "The

Olmsted planned seven increasingly formal garden areas around the house, and planted the first forest in the country to be scientifically managed. The head gardener came to Biltmore for a month and stayed a lifetime.

mountains are just the right size and scale for the château!" Herculean labors were going into the making of George's estate: 1,000 workers laboring ten hours a day, six days a week, at the peak building period (1892–95); a three-mile railroad spur built to deliver construction materials to the site; 15,322 bricks made in one morning in the estate's own kilns; a mountaintop leveled to create the platform for the house; a village in the making for the workers, to include houses, churches, schools, a hospital, and vocational training schools for local youngsters.

By the time the house was finished, it was vast enough to fit everything George had inherited, acquired, or commissioned: his thousands of books (he read in eight languages); the 300 oriental rugs bought in a single morning at a London warehouse; a painting 64 by 32 feet for the library ceiling; a dining room table that could seat 64 for a Banquet Hall 72 feet long, 42 feet wide, and 70 feet high; one of the greatest private collections of Sargent paintings in the world; portraits of family and friends by Boldini and George's good friend James McNeill Whistler; 1,600 prints by masters like Rembrandt and Dürer; Cardinal Richelieu's wall hangings; Napoleon's game table and ivory chess pieces; and a 1,700-pound chandelier, perhaps the world's largest such fixture suspended from a single point.

And that was only on the inside. Outside, Frederick Law Olmsted was realizing the plans he had presented to George back in July 1889. The powers of organization and administration he had displayed as head of the Sanitary Commission during the Civil War, and afterward as director of the Southern Famine Relief Association, were in full play here. Except for an arboretum, everything Olmsted had envisioned came into being:

> [his] plans for the forest, the park, the agricultural bottomlands, the approach road, the home grounds . . . the esplanade, the south terrace, the bowling green, the ramble, and the walled garden . . .

The scale of the operation was worthy of a pharaoh. In 1889, Olmsted had put up two observation towers to figure out the best lines and elevations for the site of the house. By 1891, he had set up four weather stations at different elevations to determine what kinds of vegetation could survive at what altitudes. He already had on site two or more specimens of 4,200 varieties of trees and shrubs. By 1893, close to 3 million plants would be in place. The secretary of agriculture of the day complained that George's forester, Gifford Pinchot, was "spending more money than Congress appropriates for this department. He employs more men than I have in my charge."

Olmsted moved the three-mile-long approach road to the house from the mountain ridge to a ravine below, as he put it, "to give it a comparatively wild and secluded character . . . consistent with the impression of passing through the remote depths of the natural forest." He altered the streambed in the ravine, so "the water should appear sometimes in still broad pools . . . sometimes in rapids . . . and sometimes in small cascades with yet bolder banks buttressed with rock." When Olmsted had finished, what looked totally natural had in fact been planned down to the smallest shrub.

The huge chandelier hangs above the main staircase, perhaps the largest in the world suspended from a single point.

But that was only part of Olmsted's dramatic vision. He had planned the approach to the house so that after a long, gently winding drive through the forest, "with an abrupt transition into the enclosure of the trim, level, open, airy, spacious thoroughly artificial court . . . the Residence, with its orderly dependencies, breaks suddenly and fully upon" the astonished visitor. It worked—the effect is still breathtaking. Hunt told George the approach road alone "would give Olmsted lasting fame."

"The Residence" was as heroic an achievement for Hunt as Olmsted's shaping of Vanderbilt's lands. Behind the French Renaissance facade is construction as thoroughly modern for its time as the design was traditional. Hunt may have seemed old-fashioned and derivative to his critics, but he was up to speed on the latest technology. Biltmore's floor joists and rafters were steel. It was one of the first homes in America to be centrally heated. Three huge boilers could heat even the third floor up to 60 degrees, luxurious warmth in those days. It had its own electricity generator. Water came from a 2.5-million-gallon reservoir at the top of a nearby mountain that George also owned. Guests got hot water the minute they turned on the tap. The passenger elevator was the first in town; the mechanized refrigeration system was a rarity in the late nineteenth century.

By Christmas of 1895, six years after construction began, Biltmore was just about ready for George's first house party. Twenty-seven Vanderbilt siblings, husbands, wives, and children arrived in private railroad cars. The national press was enthralled. George's

George was the youngest of William Vanderbilt's eight children. He sits in the red chair surrounded by the family in their Fifth Avenue mansion in 1873.

family was dumbstruck. The quiet, shy intellectual had beat them all at their own game.

George could have entertained on a lavish scale. The house was equipped like a hotel. But his guestbook lists only small groups, mainly family and close friends, with an occasional luminary, such as Edith Wharton, General Pershing, or Henry James. President McKinley dropped by while George was in Europe. How did they like it? Reactions varied. One visitor thought the house was "one long tale of delight . . . the proportion and scale, combined with the details, fill one with the kind of peace which comes from artistic perfection." Henry James, on the other hand, thought it was awful. Publicly, he said Biltmore was a "castle of enchantment . . . a modern miracle," but privately he described his own room as a "glacial phantasy . . . we measure by leagues and we sit in Cathedrals" " . . . bloated Biltmore was utterly unaddressed to any possible arrangement of life or state of society."

Something indeed was wrong. In an eerie foreshadowing of Biltmore's own fate, when he was only eighteen George had described in his diary a visit to the palace of Prince Eugene of Savoy, who died in 1736,

> so he couldn't have enjoyed his palace over 12 years. However, that seems to have been the way with most all palaces over here, they are used for a while by the owners as residences and then are left by them to be museums or confiscated during the wars. . . .

George enjoyed Biltmore longer than that; he was there off and on for nineteen years. He and Edith had canceled their booking on the *Titanic* just in time. But Hunt's hope that George would live long to enjoy his estate was not to be. He died suddenly in Washington of a heart attack after an appendectomy, in 1914. He was 51. George's noble dream of a productive, self-sustaining, model country estate had in fact begun to fade just a few years after that first Christmas party.

Some things were working very well. A rave review in the September 1902 issue of *Country Life* magazine reported success at "the Great Model Estate of George Washington Vanderbilt in North Carolina. Village, farm, dairy, forest, school, gardens, nursery, herbarium, all welded into one immense enterprise. . . ." But George had spent just about all his $5 million capital to build Biltmore. The income from his $5 million trust could not sustain the $5,000 weekly payroll for eighty house staff and hundreds more working on the estate, the quarter of a million dollars it took to maintain the landscaping projects, or the $12,670.50 electricity bill in 1906 for the house and the village. By 1902, expenses had been cut back drastically. George's wife, Edith, was seen in tears. His financial adviser was furious, at both George and Olmsted's son, who had taken over when Olmsted became too ill to continue. "The trouble with you landscape architects," he raged, "is that you don't protect your clients from their own ignorant impulsiveness." Olmsted Junior shot back, "If we had known earlier that George Vanderbilt was spending more than his income on the Biltmore Estate, and eating seriously into his capital, we could, and would, have urged methods of economizing."

George had seemed meticulous about money as a boy, and even Olmsted had described him prematurely as "shrewd, sharp, exacting and resolute in matters of business." In his diary, at age thirteen, George kept a cash account, noting his expenses and assets carefully every day. On May 27, 1875 he wrote,

The Banquet Hall was the largest room in the house and had perfect acoustics despite its size. Richard Morris Hunt was a master of both modern technology and traditional architectural styles.

> I have been way down town today and have displeased Mother, she gave me two dollars to buy a sketch book . . . but (I) could not get one so I spent it on books, besides 2.65 of my own money which Mother did not like, she said I ought to (give) back the money.

George could be very tight with money, even as he spent his millions creating the estate. James McNeill Whistler had expressed disbelief that Edith comparison-shopped restaurant prices during one European trip. The *Washington Post* reported in horror that George and Edith stayed in such a cheap hotel on another visit that they had to pass through the owner's kitchen to get to their room.

Perhaps George's diaries contain a clue to his weakness with money, what Olmsted later described as "casual and informal fiscal control." While his mother was doling out his 50-cents-a-week allowance, "Papa," it seems, was occasionally slipping him a dollar or two, on one occasion $5, all noted in George's cash account! At fifteen, his grandfather had left him $1 million, and a doting father had given him a million when he turned twenty-one. So perhaps George thought there would somehow always be more where that came from. Was Biltmore "Vanderbilt's Folly," "a devastating commentary on the injustice of concentrated wealth," as Gifford Pinchot saw it, even though he was "happy as a clam at high tide" to be working on the project? Or was it, as Olmsted believed, "a work of very rare public interest, in many ways . . . of much greater public interest, utility industrial and political, educational and otherwise, very possibly than we can define to ourselves"?

It is hard to dismiss a man as merely greedy and ostentatious whose passion for, and support of, the arts was so pronounced even in his early teens. Biltmore is Vanderbilt's most exquisite work of art, inside and out. It is hard not to feel compassion for a man remembered at his death as "courteous in manner, dignified in deportment, kind in heart and pure in morals . . . where ever there is nobility of character and gentleness of spirit . . . where ever there are all those things that make for sweetness and light, there George Vanderbilt had found his home."

And it is unfair to condemn a dreamer wise enough to choose men—Olmsted, Hunt, and Pinchot—whose combined gifts and vision created a masterpiece. Biltmore is a unique and magnificent symbol of a robust age when the United States flexed its muscles to the world as a new industrial giant that could rival Europe. We will never see anything like it again.

Today Biltmore Estate thrives under the forceful management of William A. V. Cecil, the son of George and Edith Vanderbilt's only child, Cornelia. Since 1960, Cecil has imposed on the estate a vision equal to that of the men who created Biltmore a century ago. "It is a slice of history," he insists, "and it would be the defeat of a lifetime to give it up. George may have let Biltmore get out of control," he says, "but new money, like the Vanderbilts or the Medicis, gives you great art."

Cecil has created the self-sustaining, productive enterprise of which Vanderbilt dreamed. Biltmore Estate is one of the region's largest employers and biggest taxpayers. It attracts more than 850,000 visitors a year. The house, Cecil says, is better maintained now than in George's day. It's not easy to keep it going, and Cecil worries about inheritance taxes and other economic pressures on the fate of the estate when he's gone. Nothing is certain about the future except the lasting grandeur and determination of his own dream to preserve one of the great dreams from America's past.

George was the family intellectual; this library contains 10,000 of his 23,000 books.

LOWER EAST SIDE TENEMENT MUSEUM

New York, New York

The tenement at 97 Orchard Street had been in business for three decades when this Orchard Street scene was photographed in 1898.

On a hot summer day around the turn of the century, a sixteen-year-old immigrant who'd arrived that morning in America found himself on Orchard Street, on Manhattan's Lower East Side. His first impressions were devastating:

> The crush and the stench were enough to suffocate you . . . dirty children playing in the streets . . . dark tenements and filthy sidewalks, saloons on every corner; sinister red lights in the vestibules of many small frame houses—all these shattered my illusions of America and made me feel terribly homesick for the beautiful green hills of my native Vilna.

By the end of the day, the young man had found a job and a place to live. In a flea-bitten bed, he finally got to sleep, his new life begun—and badly. One of those dark tenements on Orchard Street, No. 97, had been there since 1863, and by the time he walked by it, as many as 5,000 tenants had already come and gone.

Lucas Glockner, a tailor from Germany, built 97 Orchard Street during the Civil War. Like so many other small businessmen, he hoped to profit from the hordes of immigrants who kept pouring into New York City. Few regulations stood in his way. So on a lot even smaller than the 25-by-100-foot divisions the city intended only for single-family dwellings, Glockner put up a six-story tenement, with twenty apartments on five floors, four to a floor, each with three rooms totaling 325 square feet. The ground floor was set aside for commercial use. Despite the rudimentary housing Glockner offered, in the 1870 census 71 people were living at No. 97, including Glockner's son, Edward, his wife, Caroline, and a baby daughter, Louisa. The Tenth Ward, where 97 Orchard was located, was on its way to becoming the most crowded area in the world. Charles Dickens said it made Calcutta look like Paradise.

Glockner could have provided gas, cold running water, and stoves, but he chose not to. What his tenants got was a pump, a few

This front stoop was typical of Lower East Side tenements in 1905. Around that time, there was probably a clothing store on the street level at 97 Orchard.

privies in the small back yard, and two fireplaces per apartment for heat. If you wanted a stove, you bought it yourself. If your apartment was above the first floor, you walked up a narrow wooden staircase, and if you wanted to see where you were going, you lit a match. As for natural light and air in the kitchen and bedroom, you had to forget about that. Only one room in each apartment had windows.

No. 97 Orchard was typical of the tenements going up all over the Lower East Side. On the street side, passersby saw an inexpensive simple red brick facade in the Italianate style so commonplace in the early 1860s, a mass-produced metal cornice, with modest brownstone lintels and sills for the four front windows on each floor. The rear of the tenement was even more basic. As for fire escapes—it was best not to think about that. Glockner provided ladders—for the agile. We don't know for sure who the architect of this establishment was, but he conformed to the brutal definition the New York Superintendent of Buildings had offered in 1862: a tenement is a building "where the greatest amount of profit is sought to be realized from the least possible amount of space, with little or no regard for the health, comfort, or protection of the lives of the tenants."

A quarter of a century after the disheartened young immigrant discovered Orchard Street, Max Mason arrived in America. He was eight, and came with his mother, an older sister, and a younger brother. As soon as they found his father and uncle waiting for them on Ellis Island, they settled right in at 97 Orchard. Mason's first impressions of it were more positive, perhaps because he was so young and not alone. "The structure seemed to reach for the sky. It was a huge tower for us, because the little town that I came from in the Ukraine had small homes with thatched roofs that were maybe a floor-and-a-half high. No. 97 Orchard was a busy, busy place, and outside was the bustle of a busy world full of pushcarts and noise and crowds," he remembered when he saw the tenement again seventy years later, in 1992. But in hindsight he added, "I marvel at how anyone could have survived living here."

Whatever brought lawyer Max Mason back to dilapidated 97 Orchard Street after so many years must have been pretty compelling. A century ago, reporter Jacob Riis, whose photographs and stories had exposed horrifying slum conditions in places like

LEFT: *The Lower East Side Tenement Museum is the only museum in the country dedicated to the urban pioneer experience.*

ABOVE: *This doorjamb of an apartment at 97 Orchard Street, with perhaps a shop inventory penciled on it, is one of the museum's archaeological treasures.*

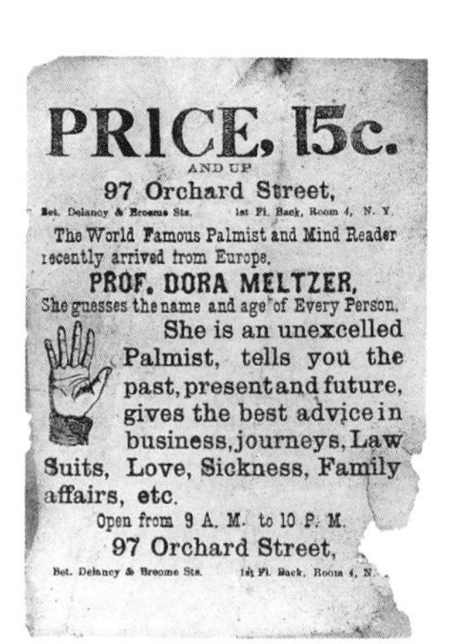

ABOVE: *Dora Meltzer was one of the 10,000 people who lived at 97 Orchard Street in its seventy-year existence.*

RIGHT: *Nathalia Gumpertz's front room became a sewing room after her husband deserted the family in 1874.*

97 Orchard, wrote in his classic *How the Other Half Lives,* "As soon as he can save up money enough, [the tenement dweller] gets out and never crosses the threshold of one again." Not only that, the tenement experience wasn't something escapees even wanted to talk about. Most of them chose to forget it as fast as they could, even to hide the fact that they'd ever lived in such a place. Somehow it seemed shameful.

Max Mason came back to 97 Orchard for a unique celebration. What neighbors used to call "the dump" had become the Lower East Side Tenement Museum, the first and only such museum in the country, a "watershed moment in the history of museums," said a Smithsonian scholar, "virtually alone in its focus on the housing and lives of urban working people." The museum is the dream come true of its founder-director, Ruth Abram. She had about given up her search in 1988 for a place to tell the story of what she calls the "urban pioneers," the millions of Americans who went west, not across the Plains, but across the ocean, to face the perils, not of isolation, Indian attacks, and wild animals, but of culture shock, overcrowding, and exploitation. Then, a colleague, Anita Jacobson, who became the museum's first curator, stumbled onto 97 Orchard. As Abram and Jacobson made their way through the boarded-up, derelict tenement, they felt like archaeologists opening a long-sealed tomb. The treasures they found were modest: scribbled inventory accounts on a doorjamb in one apartment;

layer on layer of wallpaper peeling off in others; medicine bottles, toys, and Professor Dora Meltzer's proud advertisement: "The World Famous Palmist and Mind Reader recently arrived from Europe . . . who gives the best advice in business, journeys . . . love, sickness, family affairs, etc." (All that and more for 15 cents, in the First Floor Rear apartment.) It's not the romanticized log cabin on the prairie that most of us share as a common past, Abram argues, it's tenements like 97 Orchard. Nostalgia comes easier for what you don't really know, or as one former Orchard Street resident put it, "Anybody who says those were the good old days is full of it."

The tiny bedroom in the Baldizzi apartment shows the opening onto a ventilation shaft, one of the reforms intended to bring more light and air into each apartment.

As many as 10,000 men, women, and children from twenty countries may have lived at 97 Orchard from 1863 until it was condemned in 1935. The museum staff has tracked down over a thousand of them. Census data, ship manifests, city directories, draft registrations from the Civil War and World War I, birth, marriage, and death certificates, court records, and voter registration rolls—all have turned up precious information. More and more, too, the museum is collecting memories from former tenants like Max Mason and the descendants of former residents, who are beginning to appreciate that surviving in a place like 97 Orchard took guts and vision, and that having such experience in your family background may be something to be proud of.

It helps to have a flashlight to find your way up the dark stairs and dark halls to the apartments the museum has now re-created around the lives of several families who once lived at 97 Orchard: the Gumpertzes in the 1870s, the Confinos during World War I, and the Baldizzis during the Great Depression.

Nathalia Gumpertz and her three daughters had a front apartment on the second floor. The one room with windows looked out over Orchard Street and gave her the daylight she needed for her dressmaking; kerosene lamps lit the night. Her husband, Julius, had left for work early one October morning in 1874, and never came home. It must have been small comfort for Nathalia to know that there were hundreds of women in the same predicament. Desertion was a serious, persistent problem in immigrant families.

The Baldizzi family cooked, ate, bathed, and worked in this kitchen. They left in 1935 when the tenement was boarded up.

But Nathalia didn't have to apply for relief. She became a seamstress, one of the few respectable occupations for women in the city in those days. Around the corner, Allen Street was notorious for prostitution. Nathalia had tough competition. In 1870, there were 35,000 dressmakers in New York City alone. But if she earned $7.50 or $8 a week, she could pay the rent and school costs for the children. At $10 for a new dress, or $5 for remaking an old one, her family could survive.

In 1883, Nathalia went to court to try to get her hands on a $600 inheritance she learned had been left to Julius on his father's death back in Prussia. Lucas Glockner and John Schneider, the saloonkeeper in the basement at 97 Orchard, testified on her behalf. She must have won her suit because soon after, she and the girls moved up to the German-speaking "suburb" Yorkville. She died in 1894 of a cerebral hemorrhage—at age fifty-eight.

The Confino family arrived at 97 Orchard in 1913. They had been prominent members of the Jewish community in Kastoria, Greece, but lost everything in a fire, so eventually they decided they might be better off in America. Cousin Joe was already there, safe from the Greek Army's clutches; so was their married daughter, Allegra. All eight of the Confinos and two nephews moved into one of the cheapest apartments. That meant the fifth floor, but at least it was on the front. Like all the apartments at No.

97, it had only three rooms as small, Riis said, "as if they were made for dwarfs. There wasn't room to swing the proverbial cat in any one of them."

While Abraham tried to make a living as a peddler, a terrible comedown for him, Rachel was keeping house under almost impossible conditions. Cold running water had been run into the kitchens by then, thanks to a court decision in 1895. And now each floor had two toilets in the hallway. The U.S. Supreme Court had to rule on that one! Reformers were now focusing on improving ventilation in the two inner rooms of each apartment. Inch by inch, it seemed, conditions were being corrected. Riis had revealed that one child's death in a similar tenement was reported as "plainly due to suffocation in the foul air of an unventilated apartment." So windows had been cut in the interior walls in the Confinos' apartment. Now the kitchen (132 square feet) and the bedroom (67 square feet) got some light and air. But with ten people eating, living, and often working in that tiny space, "tenement stink" would have been hard to eliminate.

By the time the Baldizzis moved in in the late twenties, 97 Orchard was about finished as residential property. In 1870, 71 people were living there. The peak, 111 tenants, was reached in 1900. By 1925, the number of tenants had fallen to 59. Rosaria and Adolfo Baldizzi came to the United States from Sicily—illegally, their daughter Josephine thought. They'd moved a number of times, probably taking advantage of the month's free rent landlords offered to induce prospective tenants into their particular slum. If you "forgot" to pay the last month's rent at your previous residence, you got away with two months for free. It couldn't have been a worse time for Adolfo. He was a skilled cabinetmaker, but business was so bad during the Great Depression, that he walked the street with his toolbox, looking for odd jobs. Eleanor Roosevelt gave him one, Josephine recalled.

Rosaria was a fanatic housekeeper. "Shine 'em up Sadie" her neighbors called her. She and "Al" used the front room as their bedroom. They put Josephine and her younger brother, John, to bed there, and moved the two onto a cot in the inner bedroom after they'd fallen asleep. Morning glories grew in their two windows. Josephine remembered the beautiful embroidered curtains, her mother's starched dresses and earrings, their pet birds, and the friendly neighbors of other faiths. But she also remembered the dim light, the cold, the daily sponge baths in the kitchen sink that were even colder. The hall toilets were smelly, the walls thin, the street still noisy, the furniture highly polished but sparse, and meat for dinner even more so. Finally, the Baldizzis had to get out of 97 Orchard. In 1935, the landlord decided to close it up rather than conform to the newest regulation: that each apartment have its own toilet. From beginning to end, there never was a bathtub, a shower, or a toilet inside any of the apartments at 97 Orchard.

The Tenement Museum staff searches constantly for "alumni" of 97 Orchard, and more and more "urban pioneers" are willing to share their most vivid recollections of tenement life. They realize, as Riis concluded long ago, that the people didn't make the slums, the slum environment made the slum dweller.

> How well I remember the only available bathtub in those days. It was in my aunt's house. After a long time, she called, "Rose, are you there?" "Yes, I'm here." That was my first bath in a bathtub and my last.
>
> It was no unusual sight to see the people sleeping all night on the roofs or the steps.
>
> The only private spot, a fire escape, where in good weather, I could read a book and perhaps catch a glimpse of the sky.
>
> My father . . . usually slept in shifts . . . not lengthwise on the bed but across it; sometimes they slept in that fashion three or four men together. What a life!
>
> The kitchen was the most lived-in room in the apartment. [Zero Mostel described the kitchen in his apartment on the Lower East Side as his own private Coney Island.]
>
> Mother's sleeping quarters was in front of the kitchen stove—the bed was made up of two chairs upon which bed clothes were piled on at night.
>
> Bedbugs did like to chew on us.
>
> Rats were not uncommon . . . to hear the feet prancing about our covers at night as we were sleeping . . .
>
> My cousin pointed out to me several of my former fellow-townspeople—men of worth and standing . . . Jonah Gershon . . . chairman of the hospital committee in Vaslui . . . dispensing soda water and selling lollypops on the corner. . . . Layvis . . . after two years' training in medicine at the University of Bucharest . . . selling newspapers on the street.

What is remarkable about 97 Orchard is the tenants' constant search for beauty amid the squalor. The inhabitants clearly tried to make a home out of their tiny spaces, and they followed the popular middle-class fashions in interior decoration and design. Experts have analyzed the paint, the linoleum, and the wallpaper in the sixty rooms at

Keeping clean in a tenement apartment was not easy. In seventy years, the landlord never provided hot water or a toilet in any of the twenty apartments.

97 Orchard. Paint must have been applied approximately every two years over the tenement's seventy-two-year life span, they figure, given the thirty-seven to thirty-nine layers of paint they've found on average in the rooms. They conclude that the landlord provided the paint, because of the consistency of vivid colors, applied even to the ceilings–clear blue, light green, salmon pink, lavender, and, later on, rose and terra cotta. A high-quality gold-leaf paint applied to a chair rail was found in one apartment. When wallpaper became affordable around 1885, at four cents a roll and up, reformers recommended against it because they thought it led to insect infestation and the accumulation of filth and germs. But the inhabitants of 97 Orchard would have none of that. The museum's wallpaper study found twenty-two layers of wallpaper in one room alone. You'd think inhabitants would have picked paper that didn't make their tiny bedrooms or front rooms look smaller, but they did, in designs with clusters of roses, arabesques, medallions, pagodas, multicolored patterns, and shimmering effects. Former tenants remember the beautiful oriental carpet design in the linoleum–anything to help dispel the gloom.

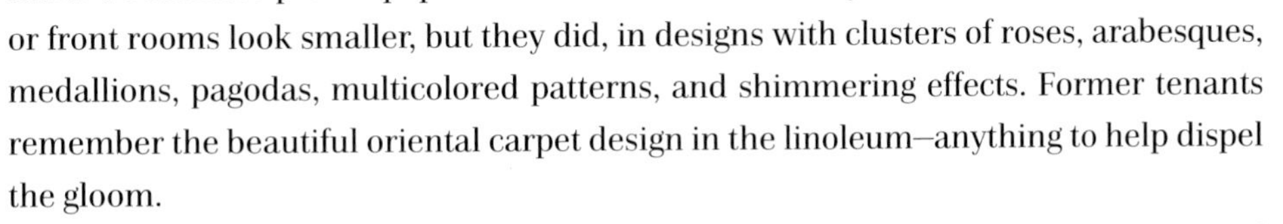

Another study came up with even more surprising data: despite the reputation of the Lower East Side as crime-ridden, 97 Orchard Street over the space of thirty years was remarkably crime-free. Only forty-five residents appeared before the Magistrate's Court between 1899 and 1930. Disorderly conduct, doing business without a license, and violation of the cigarette tax were typical of the minor offenses involved.

For every project completed, the Lower East Side Tenement Museum comes up with two or more new ones, involving the building itself and the community around it. That is because immigrants continue to pour in, to crowd into tiny apartments, work for next to nothing in sweatshops, and face old prejudices. A major mission of the Tenement Museum is to foster tolerance for their predicament and help them out of it.

For those who worry about America's future in the face of continuing immigration, 97 Orchard Street takes on an even deeper significance. Those who survived it are "proof positive that the American dream is attainable," says the museum's founder-director. "Immigrants' hopes have always recharged America's dreams. Their coming affirms us in our belief that our nation is a special place where miracles can happen. In a nation often rocked by differences our collective experience as immigrants can provide a critical common ground."

'IOLANI PALACE

HONOLULU, HAWAI'I

ABOVE: *King Kalākaua was the first monarch to travel around the world, the first to visit the United States, and the first to address the United States Congress.*

RIGHT: *In 1882, 'Iolani Palace in Honolulu became the sumptuous new residence for Hawaiian monarchs.*

WHEN THE HAWAIIAN MONARCH KING KALĀKAUA RETURNED FROM a short but historic visit to the United States in 1875, he was surprised to find that nothing was left of his royal palace but the foundations. He had been complaining for some time how unpleasant and humiliating it was for him and Queen Kapi'olani to live there. The modest coral block house had served as the royal residence for four of his predecessors, but by now, Kalākaua complained, the palace was "filthy and in poor condition," humiliating to his people as well. Sprucing it up while he was away had proved impossible; the termites had seen to that. What was left of 'Iolani Palace had to be destroyed. The only things left were its prime location and its name, Palace of the Royal Hawk.

Luckily, plans for a new palace were on hand, drawn up in 1871 during the reign of his predecessor King Kamehameha V, who had also seen the need for a more imposing residence. By 1879, King Kalākaua finally managed to get the legislature to agree to an appropriation of $50,000 for the new palace, and he laid the cornerstone on his wife's forty-fifth birthday, December 31 that year. At last, he would have a residence befitting the sovereign of an independent country recognized by the world's great powers.

Construction of the new palace was underway when King Kalākaua left in January 1881 on a daring world tour. Not only had he been the first monarch ever to visit the United States and address a joint session of Congress; he would be the first to "put a girdle around the earth," as a member of his small entourage put it. Kalākaua announced that he was going abroad to look for citizens, "so that we may become strong, secure, a nation of proud independent people—not a nation of a few millionaires and many slaves."

His sister Princess Lili'uokalani served again as regent, this time for the nine months he was away, and she was furious at the criticism of her brother's trip. Later, in her feisty, proud, caustic book, *Hawai'i's Story by Hawai'i's Queen*, she wrote,

> . . . while he was working for the prosperity of the residents of his kingdom, and for immigration which should result in the wealth of those of foreign ancestry or affiliation, they were accusing him of a reckless spending of money, and of the waste of time and revenues in foreign travels.

This wasn't the first or last stunning example of differing perceptions of the same events and personalities in Hawaiian history.

The new palace was completed in 1882. To the native Hawaiians, it was splendid, spacious, and up-to-the-minute. Kalākaua concocted an elaborate coronation ceremony for himself and Queen Kapi'olani soon after. The week-long celebration was meant to put the final seal of approval on the new dynasty he and Lili'uokalani hoped

to establish now that the Kamehameha line had died out. It was a fascinating mix of European and traditional Hawaiian royal symbolism: Hawaiian princes bore the two solid-gold royal crowns made in England, the queen and Princess Lili'uokalani wore gowns from London and Paris, Hawaiian princesses carried the royal mantle, the magnificent feather cloak of Kamehameha the Great, Hawai'i's first king. Other nobles carried the *pūlo'ulo'u* (tabu) stick and tall *kāhili* staffs.

To the non-native (haole) press, the whole thing was "cheap, tawdry and ridiculous." The king, they said, "even bungled the act of crowning himself. Instead of holding the crown with both hands, he picked it up by the small button on top, and lowered it comically onto his head. No one cheered." Lili'uokalani saw that incident very differently. "At the moment when the king was crowned," she wrote, "there appeared, shining so brilliantly as to attract great attention, a single star. It was noticed by the entire multitude . . . and a murmur of wonder and admiration passed over the throng." To Lili'uokalani, who would succeed Kalākaua as sovereign in less than a decade, the entire celebration was a smashing success. "The people . . . went back to their homes with a renewed sense of dignity and honor, involved in their nationality. . . . Naturally, those who among us did not desire to have Hawai'i remain a nation would look on an expenditure of this kind as worse than wasted. . . ."

To Kalākaua's subjects, 'Iolani Palace was a suitable reflection of his extraordinary reception by the courts of Asia and Europe during the royal passage. He was feted by the emperor of Japan, Queen Victoria of England, the maharajah of Johore, the khedive of Egypt, the kings of Siam, Italy, Belgium, and Portugal, the pope, and the president of the United States, Chester A. Arthur. (Kalākaua had dined with President Grant on his first visit to the United States.) The king made a fine impression everywhere. Mark Twain, Henry Adams, and Robert Louis Stevenson admired him. Many commented on his imposing stature (although he was well under six feet tall), his "distinguished" manners (although he was inclined to nod off if the occasion was tedious). He was charming, observant, good-humored. His two closest traveling companions took surprising liberties with that good nature. William N. Armstrong, who published *Around the World with a King* some years later, got away with telling the king that Polynesian rule in Hawai'i wasn't going to last much longer. Armstrong admitted the sagacity of Kalākaua's conclusion that the peoples of the world he had seen on his tour were no better off than his own, but of course Armstrong disagreed with the king's view that all that Hawaiians needed was to be left alone. It was far too late for that.

For centuries after the Polynesians arrived, the Hawaiian Islands had remained isolated in the North Pacific, 2,000 miles from nowhere. But once Captain James Cook happened on the island of O'ahu one January morning in 1778, to the great powers Hawai'i was now only 2,000 miles from everywhere. It became an immediate, coveted target of international power politics. The significance of Pearl Harbor, the only deepwater harbor within thousands of miles, was obvious a century before December 7, 1941. It didn't take American missionaries very long to recognize an opportunity, either; the first arrived in 1820. They introduced the alphabet, Western ways, and democratic ideals, but they damaged much that was important to the Hawaiian psyche and way of life.

America clearly was winning the international power struggle even at the time Kalākaua became king in 1874. Mark Twain, who had fallen in love with the islands on a visit in 1866, caught on fast to the extent of American influence. "Americans own the whaling fleet," he reported, "they own the great sugar plantations; they own the cattle ranches; they own their share of the mercantile depots and the lines of packet ships. Whatever of commercial and agricultural greatness the country can boast of it owes to them."

American interests in Hawai'i were also winning the struggle with the monarchy for control of the Hawaiian government itself. Too much was at stake economically and politically, as they saw it, to sit by while an independent Hawaiian monarchy went its own way, wisely or foolishly. To the Americans, the new palace was a ridiculous extravagance (the cost overruns were huge—close to $360,000), one more example of the natives' unsuitability to manage their own affairs, let alone the Americans'.

Kalākaua and his sister Lili'uokalani had seen the integrity of their own culture, and their independence, slipping away as their power base shrank with the drastic decline of the native population from perhaps 400,000 in Cook's time to 55,000 in their day. Wars, famine, disease, and demoralization had devastating consequences. They were not direct descendants of Kamehameha the Great, who had unified all the islands by the end of the eighteenth century. But they were of noble birth, and they would not go down without a battle royal. 'Iolani Palace became the grandest symbol of their fight to survive as a sovereign power and protect their culture.

To Lili'uokalani, "Kalākaua's reign was in a material sense, the golden age of Hawaiian history." 'Iolani Palace has now been restored to reflect that golden age, a time when Kalākaua dreamed of an independent Hawaiian kingdom as a key player in an Empire of the Pacific, a time when he dreamed of reviving precious expressions of traditional Hawaiian culture—the hula (so lascivious to the haoles), Hawaiian chants, Hawaiian legends. Today the palace glitters after an overhaul in the 1970s as extensive and meticulous as that of the White House during the Truman Administration.

The palace seems far too delicate and graceful a structure to have been the setting for the shattering events that came within the decade following Kalākaua's coronation, but each room in the palace also tells the story of the Hawaiian kingdom's downfall. Ironically, with its deep verandahs, high ceilings, the many windows and doors, the palace architectural style is typical of colonial government residences throughout the tropical world of the late nineteenth century. But 'Iolani was a far cry from the large grass houses where earlier monarchs held court, even if they did have Waterford chandeliers.

Portraits of the kings of the Kamehameha dynasty with their remarkable queens and chiefesses, and the popularly elected king, William Lunalilo, line the walls of the Grand Hall, which stretches from the massive front door to an identical door at the far end of the hall, both with sheet crystal panels and transom etched with the royal coat of arms. Americans dominated even that artistic enterprise. Charles Bird King, John Mix Stanley, and William Cogswell painted most of them. The massive hand-carved Grand Staircase to the royal family's private quarters on the second floor was made of koa and kamani, rare Hawaiian woods. Redwood and Port Orford cedar were imported from the Pacific Northwest for decorative woodwork. Gifts from fellow monarchs filled tall niches in the walls.

OPPOSITE TOP:
The main staircase in the Grand Hall was made of rare Hawaiian woods. Portraits of Hawai'i's kings and queens hang high on the walls.

OPPOSITE BOTTOM:
The Blue Room was the setting for triumphant and tragic events in Hawaiian history.

Kalākaua and Lili'uokalani were talented musicians, and the Blue Room, a reception room to the left of the Grand Hall on the palace's front side, was the scene of many recitals. Lili'uokalani composed hundreds of songs, including the poignant love song "Aloha 'Oe" ("Farewell to Thee"), and one Hawaiian national anthem. Kalākaua wrote the words to another. Life-size portraits of the two doomed monarchs dominate the Blue Room. A portrait of Queen Lili'uokalani's beloved husband and most trusted adviser, a New Englander and childhood friend, John Owen Dominis, also hangs there. His death less than a year after she became queen deprived her of the experienced counsel she desperately needed for the turmoil ahead. Visitors assembled here before their audiences in the Throne Room, which Lili'uokalani called the Red Chamber, or before entering the State Dining Room.

Thrones for the king and queen, and much of the furniture in the palace, were made by an American company, A. H. Davenport of Boston, in the Gothic Revival style of the late Victorian era. King Kalākaua and Queen Kapi'olani received European nobility here. Napoleon III of France and the czar of Russia, Alexander II, sent their portraits to assure the Hawaiian rulers of their appreciation of the kingdom's vital strategic position in the North Pacific. Visiting dignitaries from around the world, even A. G. Spalding and members of his baseball team, came through. The monarch's regalia, traditional and contemporary, are still displayed in the Throne Room: the coronation crowns, used just once, the royal scepter, the sword of state, and a *pūlo'ulo'u* stick (in the old days if you crossed this tabu stick deliberately, your entire family faced the fatal consequences).

King Kalākaua loved to dance (some still call him "the Merry Monarch"), and balls in the Throne Room lasted until well after midnight. The Royal Hawaiian Band, formed in 1876, still plays every Friday. An immense table dominates the king's library on the second floor, covered with huge books and examples of the correspondence Kalākaua carried on with world leaders—King Carol of Rumania, President Cleveland, and the pope, among others. Kalākaua was multilingual, and the library contains books in eleven languages, including Sanskrit. Kalākaua was interested in the latest technology, although he was an unsuccessful inventor himself. He installed the latest communications system in the palace, and spoke to the queen and friends from a telephone on his library wall. He and Thomas Edison had gotten along very well in New York, and Kalākaua, at great expense, installed electricity in 1887 with his own generating plant. Both the telephone and electricity were installed before the White House had them. Plumbing arrangements were luxurious and advanced for the time. The king's copper-lined bathtub was seven feet long, two feet wide, two feet deep. Water came from an artesian well, a new method that provided the royal family and staff with a generous supply.

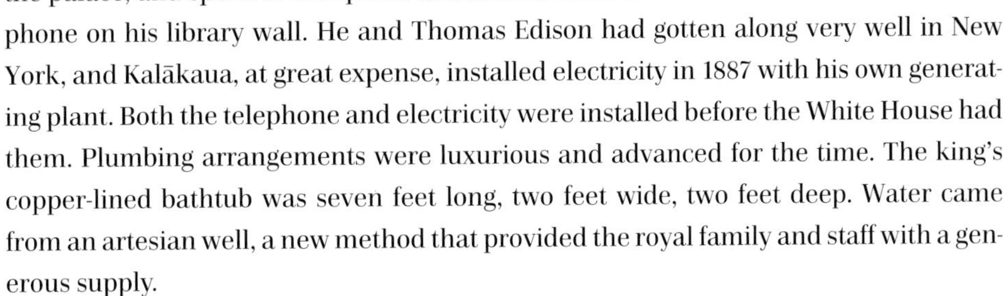

OPPOSITE: *King Kalākaua entertained lavishly in the Throne Room, but later he lay in state here, and Lili'uokalani was tried for treason here.*

ABOVE: *The Hawaiian flag was lowered for the last time on August 12, 1898, when an independent Hawaiian kingdom came to an end and Hawai'i was annexed to the United States.*

But Kalākaua enjoyed his splendid royal residence for only eight years. In the Blue Room, Kalākaua was forced to sign—under rumors of assassination if he didn't—what became known as the "Bayonet Constitution" of 1887. It left him with very little power, and the monarchy's haole opponents with practically all of it. But Kalākaua kept on fighting, and in 1889, the palace grounds were occupied briefly by a rebellious group that wanted a new native government, even if not led by Kalākaua himself. He died in 1891 in San Francisco, his heart broken, Lili'uokalani believed, "by the base ingratitude of the very persons whose fortunes he had made." Queen Kapi'olani watched from the palace's upper balcony as her husband's coffin was brought ashore. The king lay in state in the Throne Room, his coffin placed on a royal feather cloak, and surrounded by the kāhili that symbolized the royal power he once had. The staff carriers sang old-time chants of the Kalākaua family.

Lili'uokalani learned of her brother's death in the Blue Room, and there she underwent her first difficult test of queenly power with what she contemptously called the "missionary party," sons of the first American missionaries, who had gained so much economic and political power. It was a draw, but only temporarily. Two years

This huge bronze statue of Kamehameha the Great faces the palace. He unified the Hawaiian Islands and ruled from 1795 to 1819. His royal mantle was made of brightly colored feathers of indigenous birds.

later, on January 17, 1893, after a bitter struggle to restore the monarchy to full power, she was deposed by those who wanted annexation to the United States, and the Hawaiian kingdom came to an end. A century later, in 1993, the United States government, in one of the most remarkable acts in American history, apologized to native Hawaiians for its part in Queen Liliʻuokalani's overthrow. A joint resolution issued by the U.S. Senate in 1993 contained an equally stunning change in attitude: in 1842, President John Tyler had referred to the islands as "just emerging from a state of barbarism," but the 1993 document referred to the Hawaiians' "sophisticated language, culture, and religion prior to the arrival of the first Europeans in 1778." The president of the United Churches of Christ also offered an apology for its "historical complicity in the illegal overthrow of the Hawaiian monarchy." The palace was draped in black bunting in commemoration.

From the steps of the palace, the victors in this final power struggle announced the formation of the Republic of Hawaiʻi on July 4, 1894. In 1898, Hawaiʻi was annexed by the United States and became a territory. The Spanish-American War, with its actions in the Philippines, had made that long-sought, long-fought action inevitable. In 1959, Hawaiʻi became the fiftieth state, a staggering change in less than two hundred years, from an isolated traditional society to statehood. The Hawaiian Senate now sat in the State Dining Room; the House of Representatives in the Throne Room; and the governor and staff in what had been the king's bedroom and library. Restoration of the palace to its monarchical splendor began in 1969, when a new capitol building opened.

Liliʻuokalani's troubles were not over in 1883. In 1895, after a failed counterrevolution in which she claimed no part, she was arrested and tried for misprision of treason. Her trial took place in February 1895 in what had been the Throne Room. She was found guilty, fined $5,000, and sentenced to five years at hard labor. She was fifty-six. Instead, she remained imprisoned in an upstairs bedroom of the palace for eight months:

> That first night of my imprisonment was the longest night I have ever passed in my life; it seemed as though the dawn of day would never come. . . .Outside of the rooms occupied by myself and my companion there were guards stationed by day and by night. . . . The sound of their never-ceasing footsteps as they tramped on their beat fell incessantly on my ears. . . . I could not but be reminded every instant that I was a prisoner. . . .

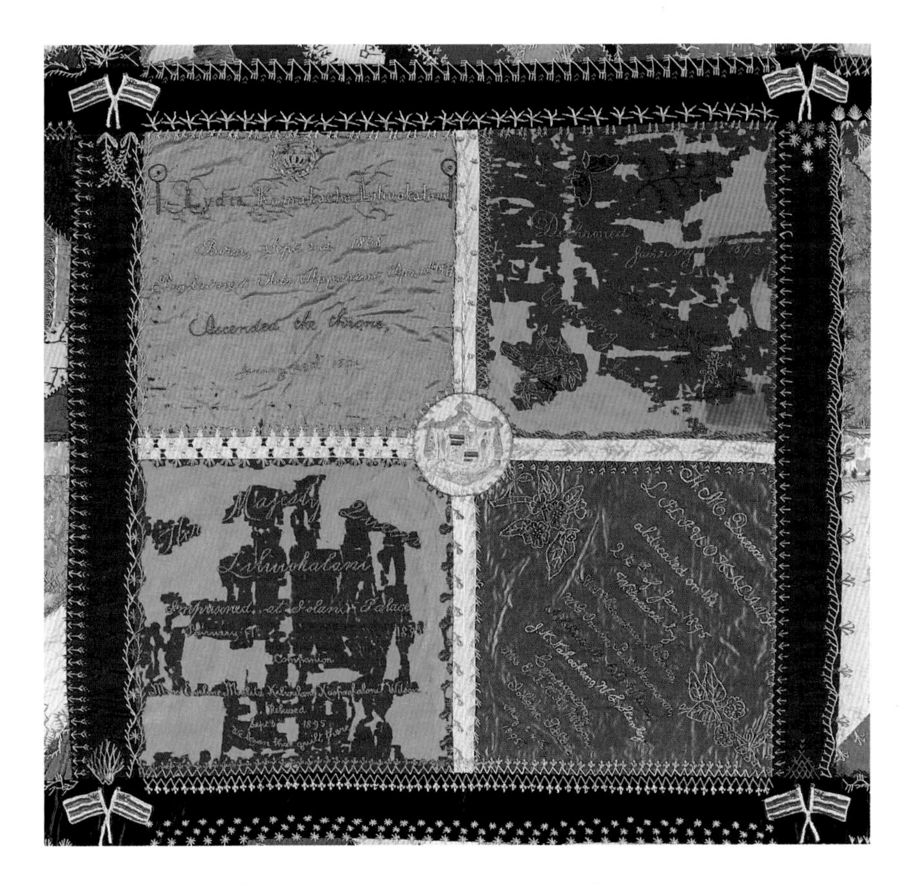

LEFT AND ABOVE:
King Kalākaua's sister, Lili'uokalani, overthrown on January 17, 1893, was the last monarch of the Hawaiian kingdom. After her trial, she embroidered signifigant events in her life on the quilt she made during her imprisonment in the Palace.

For a long time, she got her news of the outside only from the newspapers surreptiously wrapped around the flowers brought to her every day from her garden. A devout Christian, she passed the time praying, writing music, reading, practicing her guitar and ukulele. And she began a quilt on which she embroidered key events in her lifetime. This unique historical document is now the most treasured possession in the Imprisonment Room. Lili'uokalani revealed years later that she was never again comfortable "with the gaze of strangers."

Like the other houses in this chapter, 'Iolani Palace was first built on dreams: Kalākaua's dreams. Those dreams, too, succumbed to momentous shifts in power and wealth. But the palace, like the other houses we have visited, has been reshaped to fit a new dream: the rebirth of Hawai'i's traditional culture. A renaissance is underway, and 'Iolani Palace has become a focal point for the consideration of the many possibilities that await what Mark Twain devotedly called "the true Isles of the Blest."

VII

HAUNTED HOUSES

There are no conventional ghosts in these houses, no mysterious footsteps or strange voices in the night. The first, a cottage in the Bronx, was the last refuge of Edgar Allan Poe, the much maligned "genius of American terror," who was destroyed by the phantoms that haunted him all his life. Janet Sherlock Smith's hotel has seen South Pass City, Wyoming, rise from the dead numerous times. It is one of the many ghost towns in that "land of a hundred thousand phantoms." Edward Gorey's Cape Cod house looks from the outside as if it could well be haunted, but the inside reflects the unique perspective of an artist known as "the Master of the Macabre."

EDGAR ALLAN POE'S COTTAGE

BRONX, NEW YORK

RIGHT: *Edgar Allan Poe's country cottage, "the Literary Gateway to the Nation," today nestles in a small park surrounded by New York City high-rises.*

PREVIOUS SPREAD: *The cottage was surrounded in Poe's time by trees, flowers, and a beautiful lawn. He loved to garden. This 1884 photograph shows the cottage in its original spot along the east side of Kingsbridge Road.*

EDGAR ALLAN POE (A.K.A. EDGAR A. PERRY, HENRI LE RENNET, EDWARD S. T. Grey) knew precisely how the house of a proper gentleman should look. It bore no resemblance whatever to the simple country cottage where he spent the last three years of his unhappy life. Poe's "Philosophy of Furniture" appeared in the May 1840 issue of *Burton's Gentleman's Magazine.* He was thirty-one, the assistant editor, and as merciless a critic of American taste as he was of American literature."I intend to put up with nothing I cannot *put down,*" Poe boasted, and became known as "the Comanche of literature, the tomahawk man."

In the "Philosophy of Furniture," Poe launched a typical attack:

> There could be nothing more directly offensive to the eye of an artist than the interior of what is termed in the United States . . . a well-furnished apartment . . . Flickering, unquiet lights are sometimes pleasing–to children and idiots . . . no one having both brains and eyes will use [gaslight] . . . representation of well-known objects should not be endured within the limits of Christendom. . . .

Edgar had lived in luxury as a child, but when he was twenty-one, he left his foster father's mansion Moldavia for good. Poe's mother, Eliza, a talented, popular actress and singer, had died in Richmond, Virginia, during a road tour. His father, David Poe, an untalented alcoholic actor (one critic concluded "a footman is the extent of what he ought to attempt"), had already deserted her and Edgar, just three years old, his older brother, William Henry Leonard, and an infant sister, Rosalie. John and Fanny Allan, a childless Richmond couple, had taken Edgar in, and as he grew up, it became clear that he and John Allan were poles apart temperamentally and aesthetically. John Allan's first wife, Fanny, an orphan herself at ten, had mediated many bitter quarrels between the two. Allan was a prosperous merchant, but conventional and often stingy, perhaps resentful enough of his own harsh youth not to want to make life any easier for his sensitive, imaginative foster son.

Allan's mansion, with its mirrored ballroom, octagonal dining room, rich draperies, busts and paintings everywhere, could have been Poe's model in "the Philosophy of Furniture" for everything that was wrong with American home decoration. And Poe perhaps had Allan in mind when he went on, "It is an evil growing out of our republican institutions, that here a man of large purse has usually a very little soul which he keeps it in. The corruption of taste is a portion or a pendant of the dollar manufacture."

POE COTTAGE
EDGAR ALLAN POE

Allan's "little soul" had led to Edgar's having to leave the University of Virginia after only eight months, despite his fine scholastic record. Poe insisted that Allan had not given him nearly enough to cover even his basic expenses, and his attempts to make ends meet by gambling were disastrous. Allan's "little soul" prompted his refusal to give Edgar permission to withdraw from West Point, leaving Poe to provoke a court-martial deliberately as the only way out to the creative life he craved. It had led to a will in 1834 that left not a penny of his money to his foster son. For the rest of his life, Edgar could not bring himself to use Allan's full name in his signature. He was always Edgar A. Poe, and his aliases served to cover the embarrassing financial situations he inevitably found himself in.

To satisfy his desperate longing for a home and a family whose love he could count on, Poe went back to Baltimore in 1831 to live with his aunt, the resourceful, loyal, maternal Maria Poe Clemm, and her daughter, Virginia, then only nine. Poe's move to their tiny house on Amity Street made sense. He was the fourth generation of the Poe family in town; his grandfather, General David Poe, had served with distinction in the Revolutionary War and the War of 1812. "Eddie, Muddy, and Sissy"—as they called one another—would spend the rest of their lives together, in abject poverty and ill health, moving in search of a livelihood from Baltimore, to Richmond, to Philadelphia, finally to New York, and from one "old and buggy" house to another. Through everything, Poe remained devoted to Virginia and his aunt, even to their beloved cat, Catterina.

Poe's ideal room would never come remotely within his grasp, no matter how hard he worked, or how prolific and brilliant he was as an editor, literary critic, and writer. The room Poe described in his "Philosophy" was large and ornate. Poe's dismissal of it as "small and not ostentatious" was a diversionary tactic to draw attention from his own pathetic circumstances, and to remind readers of the aristocratic lifestyle he believed John Allan had denied him. The windowpanes in this room were of "crimson-tinted glass, set in massive rose-wood framings," the curtains were of "thick silver tissue," the drapes of "exceedingly rich crimson silk lined with silver tissue, closed by means of a thick rope of gold," the carpet "quite half an inch thick," the walls "prepared with a glossy paper of a silver grey tint, spotted with small Arabesque devises of a fainter hue of the prevalent crimson." On and on, Poe detailed his opulent vision: "two large sofas of rose-wood and crimson silk . . . the rose-wood pianoforte without cover and thrown open" . . . "four large and gorgeous Sevres vases, in which bloom a profusion of sweet and vivid flowers." How excruciating it must have been to have to offer a visitor to the cottage what she later described as "a small wooden box to sit on at the table instead of a chair."

Poe never earned more than a pittance during his lifetime ($435 in 1844, $699 in 1845, $166 in 1848) despite an enormous output. He published several volumes of literary criticism, scores of poems, and seventy pieces of fiction. He invented the modern detective story, and with sensational Gothic tales, he satisfied readers' relish for "the ludicrous heightened into the grotesque, the fearful coloured into the horrible, the witty exaggerated into the burlesque, the singular wrought out into the strange and mystical." Even as a boy, Poe liked to scare people.

This drawing is believed to be of Virginia Clemm, who adored her "Eddie." Said to be "angelically beautiful," she died of consumption not long after the impoverished Poes moved to the cottage.

Sometimes Poe was published but not paid. Once, in desperation, he offered a publisher all the profits from a new collection of his prose tales in exchange for twenty free copies. That one letter sold at auction a century later for $3,000, and one part of the proposed publication sold at auction in 1995 for $63,000. Poe sold "Ulalume" for just enough to buy a new pair of shoes, with twelve shillings left over. To survive as a writer in those days required an independent income, such as Longfellow had. And a Boston writer added insult to injury: "Let him remember how much of his pecuniary distress he had brought on through the indulgence of his own *weaknesses*."

Virginia Clemm had become Poe's "darling little wifey" in 1836. She was thirteen, and they adored each other. In 1846, she wrote him a Valentine's Day poem, each line beginning with a letter of his name:

> Ever with thee I wish to roam—
> Dearest my life is thine,
> Give me a cottage for my home
> And a rich old cypress vine,
> Removed from the world with its sin and care
> And the tattling of many tongues.
> Love alone shall guide us when we are there . . .

That spring, Poe decided to move the family to just such a cottage in the tiny village of Fordham, thirteen miles north of New York City. By then, Virginia was desperately ill with consumption, and Poe hoped she might get better in the country air. He had taken her out to see the one-and-a-half-story cottage, "half buried in fruit trees, which were then all in blossom. She was charmed with the little place, which was rented for a very trifling sum." Muddy was also pleased. "It was the sweetest little cottage imaginable," she said. "How supremely happy we were in our dear cottage home. We three lived only for each other. Eddie rarely left his beautiful house. I attended to his literary business for he, poor fellow, knew nothing about money transactions. How should he, brought up in luxury and extravagance. He passed the greater part of the morning in his study, and after he had finished his task for the day, he worked in our beautiful flower garden, or read and recited poetry to us. Everyone who knew him *intimately* loved him." Poe himself seemed content, too, despite his admission to Virginia in June that she was the only thing that made him able "to battle with this uncongenial, unsatisfactory and ungrateful life." "She loved to sit close to him when he wrote, keep his pens in order, and address his MS," a friend reported. "It was all she could do to help him in his work. Nor was she jealous of the many women who made fools of themselves over him."

Every visitor commented on the tidiness and cleanliness of their shingled cottage, the whole place not half as big as his one ideal room.

> [It] had an air of taste and gentility that must have been lent to it by the presence of its inmates. So neat, so poor, so unfurnished, and yet so charming a dwelling I never saw. The floor of the kitchen was as white as wheaten flour. A table, a chair, and a little stove that it contained seemed to furnish it perfectly. The sitting room was laid with check matting; four chairs, a light stand, and a hanging bookshelf completed its furniture. There were pretty presentation copies of books on the little shelves, and the Brownings had posts of honour on the stand. . . .

A young neighbor came upon a happy scene that summer. Poe was perched on a limb of a cherry tree near the cottage, tossing ripe fruit down to Virginia, all dressed in white, laughing and calling up to him. Suddenly, she began to cough up blood, her white dress spattered with it. Poe leaped down and carried her into the cottage. "They were awful poor," the girl noticed. "We knew the sadness of their lives," said another neighbor. "I recall the dying wife as a pallid brunette, slight, delicate stature, dark hair and eyes and most ethereal presence. Poe's devotion to her was never ending. His very hopelessness and his material inability to do for her countless things the supersensitive spirit and alert love a man of his mold prompted, must have been maddening to him."

When fall came, things got worse. Muddy could no longer supplement their meager diet with the greens she was seen digging up on their lawn. Even the $100-a-year rent

The Poes often had little to eat, but visitors were always impressed by the cottage's neatness and the Poes' gracious hospitality.

for the country cottage was impossible to pay. Notices appeared in East Coast newspapers about the Poes' destitution. "Mrs. Poe sank rapidly," wrote a visitor. "There was no clothing on the bed, which was only straw, but a snow white spread and sheets. . . . She lay on the straw bed, wrapped in her husband's great-coat, with a large tortoise-shell cat on her bosom. The wonderful cat seemed conscious of her great usefulness. The coat and the cat were the sufferer's only means of warmth, except as her husband held her hands, and her mother her feet. . . ."

Virginia died on January 30, 1847; she was twenty-four. Her five-year struggle with the disease that caused one-fourth of all deaths at that time had driven Poe insane, he later told friends, "with long intervals of horrible sanity . . ." "I see no one among the liv-

The sawed-off ends of two bedposts, cut to fit under the attic's sloping roof, substantiate the belief that this was Virginia's bed. She died in it on January 30, 1847. Poe died October 7, 1849.

ing as beautiful as my little wife." For months, he was distraught, haunted not only by her loss but by the earlier deaths of his idealized mother and his brother from the same disease and at Virginia's age; the loss of his affectionate foster mother, Fanny Allan; the scorn of John Allan's second wife, and what he saw as the betrayal by his foster father of promises of an education and inheritance fit for a gentleman.

Eventually, Poe thought he had recovered. With the cat now purring on his shoulder, Poe was able to write again. The rocking chair is still before the stone fireplace in the cottage's low-ceilinged sitting room, where he worked when the attic study was too cold. No silver wallpaper here, just colored lime wash on plaster. No gilt cornices, no rosewood window frames, just simple woodwork painted a bluish green. No half-inch-thick carpet, just a rag rug, bought along with some delicacies to eat and a silver-plated urn, from the $326.48 settlement of an 1846 libel suit. Ironically, the object of that suit,

Dr. Thomas Dunn English, who had maligned Poe badly, left some of the most reliable testimony to refute the charge that Poe was a drug addict. "I should, both as a physician and a man of observation, have discovered it," he wrote in his *Reminiscences of Poe.* The use of opium as a painkiller and tranquilizer, in the form of laudanum, was common in Poe's day, and he could have learned all he needed to know for his stories from Thomas De Quincey's *Confessions of an English Opium Eater.*

Some of Poe's most famous poems were written at the Fordham cottage after Virginia's death: "Ulalume," his lament for Virginia, "Annabel Lee," "The Bells." Walking up and down the cottage porch at night, or arm-in-arm with Muddy in the garden, he conceived "Eureka," a prose poem about man's place in the universe. It baffled most readers at the time, but one twentieth-century scientist sees it as a daring, imaginative explanation of the riddle of darkness, that the light from some stars hasn't yet reached us. Poe lectured brilliantly up and down the East Coast, and tried again to start his own literary magazine. But the ghosts from his past were too real, his frantic search for a woman to love him too humiliating, "Mob" too ugly, his enemies too many. He died ignominiously in Baltimore a few years later, of acute alcohol poisoning.

"He was his own enemy, it is true," said a faithful friend. One critic sees Poe's 1839 story "William Wilson," which ends "how utterly thou hast murdered thyself," as a "symbolic confessional." Poe had lied, savaged, and whined his way through much of life, and he drank himself to death. Tragically, just one or two glasses, Muddy explained, "and he was not responsible for either his words or actions." Poe admitted toward the end of his life, "It has not been in pursuit of pleasure that I have periled life and reputation and reason. It has been in the desperate attempt to escape from torturing memories, from a sense of insupportable loneliness, and a dread of some strange impending doom."

But Poe had loyal and loving admirers. "His clear and vivid perception of the beautiful constituted his conscience, and unless bereft of his senses by some poison, it was hard to make him offend his taste," said one frequent visitor to the cottage. A writer Poe had frantically tried to marry after Virginia's death praised "his devotion to his wife, his courtesy, his rare gifts as a conversationalist, his social charm, his innate rectitude." A Jesuit priest at nearby St. John's College, where Poe had sought intellectual stimulation, said, "In bearing and countenance he was extremely refined. His features were somewhat sharp and very thoughtful. He was well-informed on all matters. I always thought he was a gentleman by nature and instinct."

But Poe had chosen the worst possible person as his literary executor. Rufus Griswold hated Poe, and his cruel obituary gave a taste of his lies and forgeries that would

ABOVE: *The best-known image of Poe, the "Ultima Thule" daguerreotype, was taken in early November 1848, four days after he attempted suicide, distraught over his wife's death.*

RIGHT: *Poe's bust looks over the cottage's simply furnished living room, a far cry from the elegant Virginia home of his foster parents.*

haunt Poe's memory for over a century. Nathaniel Hawthorne was the only American writer of Poe's day to appreciate the "force and originality" of his writing; to Ralph Waldo Emerson, Poe was just "that jingle man." Alive or dead, Poe's reputation always fared better in Europe. Alfred Lord Tennyson called him the "literary glory of America." Charles Baudelaire became obsessed with Poe's life and artistry, and launched his reputation on the Continent with superb translations. The list of admirers grew longer as the years went by. "My personal debt to Poe is a heavy one," said Rudyard Kipling. "Where was the detective story until Poe breathed the breath of life into it?" asked A. Conan Doyle. Dostoevsky, Strindberg, Robert Louis Stevenson, James Joyce, Debussy, Ravel, Nietzsche, Rilke, Nabokov, even F. Scott Fitzgerald, are just a few of the many great talents who have acknowledged Poe's impact on their work. "So many things came from this lowly man. I am thankful to him," said the great Latin American poet Jorge Luis Borges. Franz Kafka believed he understood Poe: "Poe was. . . a poor devil who had no defenses against the world. . . . Imagination has fewer pitfalls than reality has. . . . I know his way of escape and his dreamer's face."

The stream of visitors to Poe's Cottage today from all over the world can see that dreamer's face in the beautiful bronze bust of Poe in the cottage's modest sitting room. Of course, some may read other things into it. After all, both Abraham Lincoln and Joseph Stalin were attracted to Poe's writings, which perhaps explains our endless fascination with this ill-fated multifaceted genius.

JANET SHERLOCK SMITH'S SOUTH PASS HOTEL

South Pass City, Wyoming

This view of Wyoming ghost town South Pass City shows the South Pass Hotel in its original colors, with restaurant and saloon next door.

The *South Pass City News* of April 26, 1871, carried this fancy advertisement:

> **SOUTH PASS HOTEL!**
> *the only first-class House in the city,*
> **COMMODIOUS ROOMS**
> **NEW FURNITURE!**
> *Table constantly supplied with*
> *Game and all other available Luxuries*
> **STAGE COACHES** *continuing the line of travel between the U.P.R.R. and South Pass arrive at, and depart from this House.*

Janet McOmie Sherlock surely saw the ad, and the hotel was impossible to miss. It was one of the few imposing buildings on South Pass Avenue, the new boomtown's half-mile-long main street: two stories high, bright red-trimmed balcony and columns on a white clapboard facade, a gabled roof rather than the usual false front concealing less impressive shacks. Janet and her husband, Richard, with their three children, Maggie (five years old), Peter (two), and Jennie (one), fifty cattle, and a herd of sheep, had come to town in 1868, the year before the hotel was built. Janet could not have known when she read the ad that in just a few years she would be running the South Pass Hotel, and that it would be her family's home, and the foundation of its livelihood, well into the twentieth century.

A rich lode of gold had been discovered nearby at Carissa Gulch, and South Pass City was exploding. Like hundreds of thousands of pioneers seeking their fortune in the West, Janet and Richard had come by way of the gently rising South Pass, just a few miles to the southeast of town, where the Oregon Trail crossed the Rocky Mountains at 7,500 feet over "Uncle Sam's backbone," the Continental Divide.

This wasn't Janet's first trip over South Pass. She could very well have been in the wagon train Mark Twain met heading west through the pass in the summer of 1861. He was on his way to a new job as reporter for the Virginia City (Nevada) *Territorial Enterprise.* Janet—just seventeen, and a Mormon convert—was on her way from Scotland to Zion, the Saints' promised land in Utah. Twain wrote:

> We overtook a Mormon emigrant train of thirty-three wagons and tramping wearily along and driving their herd of loose cows, were dozens of coarse-clad and sad-looking men, women and children, who had walked as they were walking now, day after day for eight lingering weeks, and in

THE EXCHANGE.

> that time had compassed the distance our stage had come in *eight days and three hours*—seven hundred and ninety-eight miles! They were dusty and uncombed, hatless, bonnetless and ragged, and they did look so tired!

Janet had started walking in Nebraska, with her brother John, her sister-in-law, and their infant daughter. At South Pass, they had 400 miles to go, and a terrible stretch lay ahead. Everything Janet owned, including her precious books, was in one small wooden chest. Fifty pounds was the limit you could take with you for free. The man she would marry a year later may have been in that wagon train, too. Like Janet, Richard Sherlock was a convert; he'd emigrated in 1853 from Ormskirk, England, so he might even have been a driver of one of the covered wagons that served as home to the emigrants during their trek. The Mormons' well-organized "down and back" wagon trains brought converts like Janet and Richard from the East to Zion, delivering one load at Salt Lake City, then heading back East for another—624 wagons with 3,900 men, women, and children in that one summer of 1861. The Mormons even arranged the journey across the Atlantic for converts who wanted to emigrate.

By 1873, Janet was in a desperate situation. Richard had died in March, and she was left in South Pass City, a twenty-nine-year-old widow with five children now. What was worse, the town was on its way to becoming a ghost town. The population had plummeted from 3,000 during the gold rush to a few hundred by 1873. Just four families would be left by 1882. South Pass City had already lasted longer than many boomtowns; Bullfrog, Nevada, had gone bust in just a few months. But that was small consolation for Janet, and it is hard to see what held her there. Perhaps she was tired of moving or the prospect of sacrificing the fine reputation she'd earned. It could have been the company of another strong woman in town, Esther Morris, a 200-pound six-footer, the first woman in the United States to hold public office—justice of the peace. Calamity Jane may still have been in South Pass City, too.

Most important of all to Janet, the South Pass Hotel was for rent, and running a hotel was one of the few respectable ways a woman could try to make a living in those days. The hotel had been vacant for a year, not the most encouraging sign. But Janet saw that downstairs, in addition to the hotel dining room, kitchen, and office, there were two bedrooms, which her family could use. Upstairs, nine small rooms on either side of a narrow hallway, with one closet at the end of it, were ready for any customers who might show up. As soon as they were old enough, her children could work in the kitchen and dining room, make the beds, empty the chamber pots, fetch water from Willow Creek, and chop wood for the four stoves that kept the place from freezing during the harsh

A covered wagon was home to thousands of pioneers like Janet for months during their treks west. Most of the pioneers walked as much as a thousand miles from their embarkation point to keep the wagon load light.

Janet Sherlock Smith, pioneer mother and businesswoman, ran the hotel that became home to her family in 1873. Her descendants stayed on in the ghost town until 1948.

winters, when the temperature could go down to 30 below, and ten feet of snow could still be on the ground in June.

The hotel was certainly a better place for her children to live than the crude log cabin the family had moved into when they first arrived in town; even that was an improvement over the dugouts some of the miners used for shelter in the early days. It looked as if she would have little competition for business. The City Hotel across the street was filthy; it may have been a brothel. If worse came to worst and there was no school, she could teach her children to read and write with the books she had brought from Scotland. Sir Walter Scott was her favorite, but she also loved Shakespeare and Burns. She would spend $5 on a book for the family, even if that meant scrimping on something else.

So Janet decided to lease the hotel, moved the family in, and began operations in late 1873. Like everyone on that remote and cruel frontier, Janet had learned to be versatile, in case things really got tough. She was much like the South Pass City official Mark Twain met in 1861 at the town's original site down Slaughterhouse Gulch, before Indians burned it to the ground. "Think of hotel-keeper, postmaster, blacksmith, mayor, constable, city marshal and principal citizen all crammed into one skin," Twain wrote. "If he were to die all over, it would be a frightful loss to the community." When Richard was alive, he and Janet had tried running a toll bridge over the Sweetwater River, then a bathhouse, even a saloon. Nothing had worked out. But her office on the northeast corner of the hotel soon became the post office, and she the postmistress. Then Janet stocked some groceries and dry goods in the office to sell to anyone coming by for mail, the beginning of a general store the family eventually built across the street.

Janet managed to hold on to the contracts with the stage and mail companies that still ran through town as South Pass City became more isolated. The new transcontinental railroad ran farther south, and South Pass winters were too harsh for most people. This was a lucrative but exhausting business for the family: "We have had from 2 to 15 passengers on the coach besides the drivers all winter and have had to get up every morning at five o'clock and up till twelve at night," wrote Janet's oldest daughter, Maggie, in March 1882, to her brother Peter, away at school in Omaha by then. Janet charged $2 a day for boarding the stage drivers and stock tender, and 25 cents a meal for other guests. This went up to 50 cents later, when Janet built a new restaurant next door. Guests ate family-style in the hotel dining room at fixed hours.

Janet stands on the hotel balcony in July 1888, with her second husband, Jim, and several children. Robert Todd Lincoln and General Philip Sheridan stayed in the bridal suite on their way west.

Somehow, in just one year, Janet had managed to scrape together $200 to buy the hotel, and when the next gold rush boom came to South Pass City in the mid-1880s, she and the children were ready. "We have been busy cleaning house," Janet wrote Peter in May 1882. "We have got the dining room painted and repapered." Typically for the time, the colors they used were vivid, both inside and outside, not the drab, dusty shades evoked in western movies. On a walk down South Pass Avenue in good times, you could pass one saloon painted red, white, and blue, another lime green and dark green. White buildings had green trim, or red and Prussian blue. Robin's-egg blue and mustard yellow were popular colors for the muslin that covered walls and ceilings to keep down dust in log cabins. In Janet's hotel, the guest rooms were decorated with wallpaper, glued right onto the boards, and the wood trim was false-grained to look like walnut.

Compared to hotels in larger, richer mining towns—with fifty, sometimes ninety rooms, red carpets and chandeliers—everything in the South Pass Hotel was cramped and rough, but respectable enough for Secretary of War Robert Todd Lincoln, Abraham Lincoln's only surviving son, and General Philip Sheridan to stop there on their way to Yellowstone in July 1882. "They make as much fuss as if they were the Emperor of Russia," Janet wrote Peter. Naturally, Lincoln and Sheridan were put in the best room upstairs, the Bridal Suite, 10 by 12 feet, the only room with a wooden bedstead, a hand-

South Pass City's primitive town jail and sometime schoolhouse is on the extreme left in this 1906 view. The false-fronted building in center right is the Smith-Sherlock store, built in 1896, another family enterprise.

some dresser, and one of the few rooms with a wood stove. The rest of the patrons got iron bedsteads and colder rooms. When business was good, they sometimes had to sleep two to a bed, even on the floors and in the hallways, and Janet's children could always give up their own beds in a pinch. The schoolteacher, when there was one, had the room at the end of the hall in the southwest corner. It isn't known which room Butch Cassidy occupied. Despite his reputation, "Uncle Butch" was popular with the young people of South Pass City for pitching coins out to them from the saloon Janet had added to her growing list of enterprises. Alcohol and gambling may have been against her religious principles, but selling whiskey–one to four gallons a day–was a sure source of income she could not afford to lose. Even so, Janet sometimes could serve only bread and milk to her children for supper.

South Pass City wasn't as violent as nearby mining towns like Atlantic City and Miner's Delight, but it had its share. Janet's brother George was shot in 1873 during a card game. Janet stopped the lynching of his killer when she pleaded that losing her brother was bad enough without having another death on her conscience. Drunken cowboys from the trail over South Pass took offense easily and dangerously. Hunger compounded the stress, and Indian scares added enormously to the tension as late as 1906. Janet and her family had to hide more than once across the street in a massive

Maggie, Janet's oldest daughter, died of gangrene in this bedroom in February 1883. She had been trapped several days in a blizzard on her way to school in Salt Lake City.

cave that had no light or ventilation. There were false alarms. One resident woke in the middle of the night, terrified by the feather she saw at the window—an Indian, surely. It turned out to have come from a pillow stuffed into a hole in the windowpane.

But it was nature's violence that brought the worst tragedies to Janet's family. She had vaccinated several of her children against smallpox, and lost none of her brood to the diphtheria or typhoid that ravaged other families. But Janet's oldest child, the beautiful and bright Maggie, died of gangrene in 1883 upstairs in the room where Lincoln and Sheridan had slept. The sleigh carrying her to school in Salt Lake City had been trapped in a fierce blizzard. "We got her home last Sunday," Janet wrote Peter on February 17, 1883. "Her feet are frozen above her ankles and her hands above the wrist . . . she was very nearly gone when she was found. The people at Dry Sandy [stage station] had to cut her clothing to get her out of the sleigh. . . ." A few days later, Janet gave her son the tragic news. "Maggie died this Morning about ½ past two o'clock. I tell you Peter, she was brave.

About the last words she said to me was Mama I am not afraid. . . ." A few weeks later, Janet wrote again. "We were so happy Peter. Seemed to me too happy. . . . In the evening we would all come into my room. Jim too. Maggie always read aloud. She read *David Copperfield* and had got about half way through *Uncle Tom's Cabin.* Now everything has changed. This Clamity has cast a gloom over everyone."

Peter, too, met with a terrible accident a few years later, but he survived. As "the best and the smartest one in the family," Maggie said, he was the first to be sent away for the education Janet valued above all. Janet's grand hopes for him were destroyed when he was blinded in a mining accident during a summer vacation. For the rest of his life, Peter manned the family's general store, recognizing customers by the sound of their footsteps. He died in South Pass City in 1947. The "Jim too" in Janet's lament for Maggie was her second husband, James Smith, an Irish adventurer who had come to South Pass City about the same time Janet had. He'd sailed as cabin boy on merchant ships, fought with the U.S. Navy during the Mexican War, been wounded, flogged, and court-martialed, gone to California in the 1849 gold rush, and ended up in South Pass City looking for more opportunities. He found Janet, and for better or worse, she married him on June 8, 1875. Anna, their first child, was born in the hotel in 1876; Ernest, their second child, Janet's last, was born in 1880. Jim was a caring but harsh stepfather, and a hard drinker always trying to quit. "There is one thing I thank the Lord for . . . Jim is still abstaining from liquor. It makes such a difference in him," Janet wrote to Peter in June 1882. Her older boys, Will and John, left the hotel as soon as they could, but didn't leave town. The girls were furious when Jim wouldn't let seventeen musicians pay their bill after a night's lodging and breakfast in June 1894. If he had worked as hard as they had, the girls wouldn't have minded so much. The Rock Springs Band had came to town as part of a booster campaign to interest potential investors in the local gold mines. There were no takers.

In the spring after Maggie's death, Janet had written Peter: "I'm getting tired of being here. It seems as if we could make a living in a pleasanter place where the children could be educated without sending them from home." But a quarter of a century later, she was still in South Pass City, applying once again for Jim Smith's widow's pension:

> I have lived here since 1868 and have lived in the same house I am now living in since 1873. have raised all my childern here . . . and am known by most of the people living here . . . when a person gets to be in there 63 year and has worked hard all there life raising a faimly aspecially in the "Rocky Mountains" they are not good for much hard work at my age.

Janet lived another seventeen years, surviving several more booms; she died in the hotel on October 3, 1923.

Annie Smith, shown here with brother Ernest and pet antelope in July 1888, is called the first curator of South Pass City. At age seven, she began to save clothes, letters, photographs, even furniture, making this ghost town perhaps the best documented in the West.

Janet had always thought South Pass City would eventually disappear. It could have become just another ghost town, one of thousands in the West that have disappeared without a trace. But thanks to Janet's youngest daughter, Anna, it didn't. Anna may have seemed spoiled and lazy, but it was her vision and persistence that saved thousands of letters, pictures, clothes, quilts, and furniture, even Janet's books and emigrant chest—30,000 artifacts in all, making South Pass City perhaps the most authentically furnished and best-documented restored mining town in the West. At South Pass City Historic Site today, they call Anna the city's "first curator." Her picture hangs over the door to the archives room at the Visitors' Center, which over time was once a dance hall, theater, community center, and Sunday school, just as the front room of the primitive 1870 jailhouse once served as a school. Beautifully written letters of the alphabet wind around the room, but there were more interesting samples of writing for the students on the cell doors: "You will get yours in a day or two, woman!" "Oh hasten the day of our trial, Or some generous man go our bail, if I only had an old file, I'd cut out of this damned old gaol."

Janet's hotel is now restored to its jaunty self as of the last decade of the nineteenth century, during a final boom. Most of the town has decayed or disappeared, making it easier to understand why South Pass country has been called "the Land of Half a Million Phantoms." For Janet and her family aren't the only ones whose spirits are so palpable there. Tepee rings overlooking the town's abandoned mines mark the passage of others who've been in the area over 9,000 years. Some pioneers left their names on Independence Rock, not far from South Pass, but more left their bones along the Oregon Trail itself, the longest cemetery in the United States, some call it.

Many, like Norman Smith, Janet's grandson, defied the odds. Around the turn of the century, it looked as if he were stillborn, so he was wrapped up and shoved out of the way in the still-warm oven of the log cabin. When someone returned with a load of firewood and slammed the cabin door, Norman began to cry, as if he'd been slapped on the bottom, and was quickly removed from the oven. Later, as a boy, Norman left his name printed boldly on a barn door in town. It's still there, a touching affirmation of life. Eventually, he went on to a career in the United States Navy. Rear Admiral Norman Smith is now buried in Arlington National Cemetery.

EDWARD GOREY'S HOUSE

YARMOUTH PORT, MASSACHUSETTS

ABOVE: The Gashlycrumb Tinies *is one of Edward Gorey's most popular books.*

RIGHT: *"Master of the Macabre" Edward Gorey's weathered Cape Cod house stands out on the Yarmouth Port village green.*

EDWARD GOREY, "MASTER OF THE MACABRE" TO AVID FANS HERE AND abroad, is a prolific writer, illustrator, set and costume designer and collector, who has lived over a decade in Yarmouth Port, Massachusetts, a Cape Cod village settled in 1638. The first white man to build there was *Mayflower* Pilgrim Stephen Hopkins, who is said to be the only American ever mentioned in a Shakespeare play—*The Tempest*'s Stephano.

That literary link is not what led Mr. Gorey to move to Yarmouth Port after years as a leader of a counterculture elite in New York City. But he has settled down with a vengeance; indeed, he never leaves. The cottage he bought in 1985 was built in 1800 and may have started out as a half Cape, he speculates, but it is now a full Cape with additions, and that is putting it mildly. If you didn't know who lived in the house, its exterior would tell you that whoever does must be an original. It faces a lovely village green, with other houses that are trim and white with manicured lawns. "Trim" and "manicured" hardly apply to Mr. Gorey's weathered gray-shingled house and grounds: "decaying," "peeling," "dilapidated" seem more appropriate. "The Fall of the House of Usher" also comes to mind. Mr. Gorey, in fact, says it might be fun to do Poe in a different way than anyone else ever has. He could, for his work, like Poe's, is usually described as spooky, haunting, macabre, even ghoulish. Mr. Gorey doesn't look on his work as macabre, because, he explains, he doesn't linger over the violence. "Life is basically ghastly," he feels, "and maybe I'm trying to get that discomforting element across. But I also try to show that life is so absurd it's comical."

The Gorey house was built by a Captain Hawes, who was lost at sea not long after, a tragic fate such as those that befall so many Gorey characters, even the children he finds so "useful for satire and parody." (One writer observes, "In Gorey's world the infant mortality rate is higher than it was in 1556.") So the visitor is prepared for anything on entering the house, guided by Mr. Gorey to the back door since the front steps aren't easy to negotiate, given the creaking wood and overgrown bushes.

The kitchen is the only bearable room in the house, Mr. Gorey explains; why so isn't immediately apparent. This room is all white, spacious and immaculate, in part perhaps because Mr. Gorey regularly eats both breakfast and lunch at Jack's Outback, an unpretentious restaurant close by. The New England light Mr. Gorey finds so agreeable floods the room through large shining paned windows.

Jane appears, one of the many cats he has shared his houses with since 1934, except for Army service and at Harvard. "It's impossible to know what's going on in

ABOVE: *Gorey is an award-winning and prolific author, illustrator, and set and costume designer, with an international following.*

RIGHT: *The Gorey workroom is as tiny and meticulous as the images he creates here.*

Edward Gorey has amassed huge and varied collections. Thousands of books are piled on shelves, floors, chairs, and tables.

their little noggins," he says, and they appear sometimes in his enigmatic work. Mr. Gorey plays with Jane as if she were as boneless as a favorite character, Figbash.

Disarmed by the erudite Mr. Gorey's conversation, the black cat's antics, and the sparkling kitchen, what follows in the rest of the house is an astonishing surprise. There really *is* no place to sit down except in the kitchen. Every inch is covered by Gorey's collections. That includes the floors, the chairs, sofas, tables, walls, mantelpieces, even the door and window frames. There must be at least 10,000 books downstairs: on shelves, in high stacks, in boxes. There is a stash of hundreds of CDs and videotapes as well.

"There is an order here," Mr. Gorey insists. Detective stories and the supernatural are housed in one room, American literature in another, for example, and he claims to know what is in every box. Gorey characterizes his reading tastes as bizarre: Greek mythology, "Roman stuff," Japanese, Chinese, and French literature, French symbolist poetry, British satire, odd travel books, Asian ceramics. Flaubert and Agatha Christie are favorites, Henry James intensely disliked. Unknowns are admired, such as Mrs. W. K. Clifford and E. H. W. Meyerstein.

The house clearly reveals that Mr. Gorey follows his own advice to "buy it if you like it." The results of his constant attendance at local flea markets and garage sales hang or perch on whatever space is left: finials, ginger jars, trunks full of iron utensils,

The best-known Gorey image is the animated opening for the PBS Mystery! *series (detail).*

antique stuffed animals, rocks, nineteenth-century sand paintings, postcards of dead babies, even a box containing shards of green Sandwich glass, a bargain at 50 cents. His fine art collection, discreetly placed to accommodate his height and daily routine, is something else: a Berthe Morisot etching, a Bonnard drawing, ten Atget photographs, a Munch lithograph, a Klee etching, Balthus and Burchfield drawings, and illustrations by George Herriman for *archy and mehitabel.* Gorey works upstairs in a tiny room about 6 by 8 feet with one narrow window. His famously intricate drawings—mostly black and white, pen and ink—are done very small, the same size as they appear on the pages, "so I don't need much room to do this in," he explains. "I would rather not have any new ideas; unfortunately, I have more than I ever did before." His books and illustrations have won international recognition over the years. He received a Tony award for the costumes he designed for the Broadway production *Dracula,* and he did the sets as well.

Viewers of the PBS *Mystery!* series see Gorey's work every week in the animated opening. Aficionados recognize the anagrams of his name in various works, twenty-one at last count, Mrs. Regera Dowdy and D. Awdrey Gore being favorites. The license plate on his modest car reads OGDRED (no room for the Weary). Mr. Gorey finds the alphabet "a great way to organize your material. Of course eventually you run up against *X* and *Z,* and so forth and just splash around." He's done six alphabet books so far, and the temptation to describe his house that way is too great to resist:

WARD GREY ODE
(with apologies to
The Glorious Nosebleed: Fifth Alphabet)

We approached the ancient gray house ***A****nxiously*
It stood out on the quaint village green ***B****izarrely*
Brambles for lawn, crumbling steps to climb ***C****autiously*
We'd come to an odd place ***D****ecidedly.*
Round the corner turned Ogdred ***E****nigmatically*
Thus preventing our falling through ***F****atally*
Out back to the kitchen he ushered us ***G****ingerly*
So we feared it might look just as ***H****orribly.*
Jane the cat eyed the visitors ***I****nscrutably*
Mr. Weary answered each query ***J****ovially*
We then tore through the cottage ***K****inetically*
Rambling rooms—white—bright—mirrored ***L****ucidly.*
We'd expected gloom, bats, and ghosts ***M****acabrely*
Instead, windows shone, floors gleamed ***N****oticeably*
On sofas, floors, even tables—but ***O****rderly*
Books piled high, by subject ***P****olymathically.*
True, poison ivy was creeping inside ***Q****uietly*
Black cat #7 fled ***R****aucously*
Cobwebs spun in the corners ***S****urreptitiously*
Frogs clung to window frames ***T****enaciously.*
Clearly Mr. Weary collects ***U****nabashedly*
Writes, draws, reads, listens ***V****oraciously*
A hundred-odd books, set designs, costumes ***W****inningly*
*His brilliance still shines e****X****traordinarily*
Well along now, white-bearded but ***Y****outhfully*
He creates, collects, pets his cats ***Z****estfully.*

Several Gorey aphorisms are worth repeating:
Beware of this and that.
When in doubt, twirl.
Blessed are the nonchalant.

Bottles, ginger jars, sand paintings, countless CDs, videos, and other objects, including seven cats, share the house with Mr. Gorey.

THE NEW YORKER
Condé Nast TRAVELER
August 1992
VOGUE
AUGUST 1992

VIII

PLAIN AND FANTASY

The three houses in this chapter demonstrate the vitality and variety of American taste: MGM Studios' powerful set designer Cedric Gibbons shaped moviegoers' fantasy worlds for a quarter century through good times and bad. His sleek Art Moderne home was a fantastic reality. Robert Venturi's small house for his mother became one of the most important buildings of the twentieth century. Its deceptively plain facade masks a structure as complex and contradictory as American society itself. The "House for the Next Millennium" may be more than a fantasy, for its two young creators are working to make it a reality in the twenty-first century.

CEDRIC GIBBONS/ DOLORES DEL RIO HOUSE

SANTA MONICA, CALIFORNIA

"THERE'S NO PLACE LIKE HOME," DOROTHY MURMURED AS she clicked the heels of her ruby slippers together three times, and found herself magically transported from the Land of Oz back to Auntie Em's plain old house on the Kansas prairie. A key creative power behind the making of that film fantasy was Cedric Gibbons, MGM's supervising art director from 1925 to 1956. Gibbons drove himself home in a white Duesenberg, to his own dazzling creation Westfront, an Art Moderne classic that epitomizes the glamorous world he created for moviegoers for a quarter century.

Film director Billy Wilder once explained why he admired Cedric Gibbons. "He knew what to put on the screen. When he built a house for himself, he also knew what would be comfortable, simple, and yet quite remarkable." That separated Gibbons from most of the movie stars and moguls in Hollywood in the late twenties and early thirties. "They could never quite make up their minds," Wilder said. "Should they build a New England farmhouse? A Riviera villa? or a Mexican hacienda? What they liked was all kind of wildly exaggerated, and the poor architects did not know which way to go." Most movie people were not as sophisticated as Cedric Gibbons. Samuel Goldwyn had been a glove salesman, Adolph Zukor a furrier. Rudolph Valentino was a poor immigrant from Italy. Chaplin came from the slums of London. Theda Bara was not the daughter of an Arabian princess, "born in the shadow of the pyramids,"as studio publicity claimed; her father was a Cincinnati tailor. Comedian Harold Lloyd and his wife spent hours wistfully riding up and down in the elevator of their overpowering $2 million dream palace, because, Mrs. Lloyd told gossip columnist Hedda Hopper, "it was the only cozy place in the house."

OPPOSITE: *The front entrance of this Art Moderne classic gives the first hint of the setbacks that are also a dominant feature inside.*

PREVIOUS SPREAD: *The tennis court pavilion is an Art Deco jewel.*

Gibbons was probably the only Hollywood designer who'd gone to Paris in 1925 to the important Exposition Internationale des Arts Décoratifs et Industriels Modernes. The impact on his own work was immediate, and technology was creating new film stock, lighting techniques, and materials that he exploited fully. Joan Crawford, Jean Harlow, Clark Gable, William Powell, and other stars began appearing in Gibbons's trademark, the "Big White Set." Their penthouses, nightclubs, and executive suites shouted wealth, luxury, and success to moviegoers.

Depression era audiences could dream of living someday themselves in a gleaming apartment like Greta Garbo's in *Susan Lenox: Her Fall and Rise,* or finding romance in a *Grand Hotel,* but for most moviegoers, these were fantasy worlds far removed from their own bleak realities. By 1939, 85,000,000 Americans were escaping to the movies every

week. What the stars wore, and the sets where they played their parts, were as important then as the plots. Next to stars' salaries, the biggest percentage of a film's budget went to costume and set design, and at MGM, Cedric Gibbons had the power and staff to shape both the budget and the image. "Nothing, absolutely nothing, went through unless Gibbons had OK'd it," said an associate. Gibbons was handsome and dashing enough to be a star himself. But he wasn't interested in working in front of the camera. His sets were costars, anyway, and sometimes they even upstaged the actors. With his name in a film's credits, moviegoers could be sure of the latest in glamour and elegance.

LEFT: *The rear view of the house, called Westfront, reveals its Art Moderne lines.*

ABOVE: *The glamorous actress Dolores Del Rio and her husband, Cedric Gibbons, MGM's award-winning art director from 1925 to 1957, relax in the living room.*

In August 1931, a movie magazine described Gibbons's new house as "Hollywood's most amazing home—modernistic in the extreme, a forerunner in a severe simple style of architecture that is rapidly becoming possible." It is still as glamorous and elegant as his sets were. The east facade of the house is a deceptively simple, almost windowless series of stepped blocks, rather like the exterior of a soundstage. It's made of gunite, "a stucco that went to college," and of copper sheeting painted celadon green. The front door, placed off center in a series of seven shallow setbacks, is made of Monel, a metal alloy of stainless steel and copper, light and strong enough for propeller blades, and one of the new materials Gibbons used throughout the house. Two wide, low steps of terrazzo, a labor-intensive, marble-based composite, lead to the front door, which opens into a dramatic, light, spacious interior worthy of the Oscar statuette Gibbons also designed and kept winning for "Best Art Direction."

Gibbons's wife in 1930 was the Mexican actress Dolores Del Rio. Errol Flynn, who was cast as Robin Hood after an archery contest at the house, thought she was one of the most beautiful women he had ever seen, "like an Aztec princess of olive skin with wondrous eyes. She was very artistically oriented." Miss Del Rio had told friends she was "aching to play the sophisticated roles of modern drama—after having been French, Russian, Indian, Gypsy, Spanish—everything but a 'Modern!'" Gibbons's house provided a perfect setting for her to practice the part. She liked to say that her sleek black hair had inspired her husband's choice of textures and surfaces for the interior. The floors are still jet black high-tech battleship linoleum a quarter-inch thick. The walls were ivory; the pillows, drapes, and upholstery soft lemon yellow silk and velour.

The owners today prefer shades of white on the walls. Their contemporary art collection provides the splashes of color Gibbons achieved with a Van Gogh above the reception room's poured-terrazzo fireplace, the zebra rug Gary Cooper brought back from an African hunting trip, vases of flowers everywhere, many books in low gray bookcases, and perhaps an Oscar or two on the stainless-steel mantelpiece in the living room, or the Bakelite and nickel-plated steel mantelpiece in the reception room.

Staircases were dramatic devices in many films of the early thirties. Gibbons included a grand version in his Versailles sets for the film *Marie Antoinette*, even though King Louis didn't have one in the original. ("Whatever you put there, they'll believe it," Gibbons was told when he complained about a script that had Paris located near the ocean.) The tan terrazzo staircase in Gibbons's house, with its steel and nickel-plated banister, sweeps theatrically across the windowed west wall of the reception room up to a living room of "heroic proportions," as the owners describe that 25-by-45-foot space. Gibbons could have been describing his house when he wrote the entry "Motion Picture Sets" for the 1929 *Encyclopaedia Britannica*: "The more frequently the wall is broken, either with jogs or recesses, the more interesting it becomes photographically, as it gives opportunity for light and shadow." Gibbons exploited that idea on the ceilings, walls, and door frames. The underside of his grand west-front staircase forms part of the dining room ceiling, each setback glowing with hidden light, a stunning variation on the same theme.

The current owners have restored the house to perfection, and love the "heroic proportions" of their living room.

Despite the Puritan morality Hollywood had to reckon with in the thirties, bathrooms and bedrooms in films managed to convey considerable sensuality, even decadence. Director Cecil B. De Mille dared to say, "Come see your favorite actors committing your favorite sins." When some worthy charity ladies visited his own house, they were

The dramatic terrazzo staircase with steel and nickel-plated banister sweeps across Westfront from the reception area up to the living room.

disappointed, he admitted, to find the bathroom "just a plain comfortable standard American bathroom, without a square inch of onyx or ermine, without even a tap over the tub for rose water or milk." Charlie Chaplin at one time had an open pipe with a cup in his! They would not have been disappointed at Cedric Gibbons's house. The mirrored, hand-carved black marble bathroom in Dolores Del Rio's suite is as titillating today as the bath Gibbons designed for his 1929 extravaganza *Dynamite.* The heads of the screws holding the mirrors to the wall are shaped like stars. Miss Del Rio took her milk baths there, and slept twenty hours at a stretch in the small master bedroom. A trapdoor in her silverleafed dressing room led from her husband's suite below.

It is difficult even today to separate this very real house from the fantasy world Gibbons created with his film set designs. Many features in the house—the lustrous built-in banquettes, the sculptured fireplace blocks, the lighting fixtures—would have been recognized immediately by 1939 filmgoers. A pyramid-based glass-topped cocktail table seems, in fact, to have made its way back and forth from the Gibbons house to his sets. The house has served as a location for recent movies; the fireplace tools in the living room were left by one film crew.

Gibbons used the newest materials, such as Monel and Bakelite, to create sleek, sophisticated effects, at home and on his sets. This was a high-tech house in 1930.

Gibbons ended his essay in the *Encyclopaedia Britannica* with an impossible premise. "If realism can be abandoned," he wrote, "we may look for a setting which in itself will be as completely modern as is modern painting or sculpture." But realism couldn't be abandoned, not after the atom bomb and Auschwitz. Moviegoers would seek escape through different cinematic fantasies: science fiction and adventure, and, again, technological developments would aid the filmmaker, as they had Cedric Gibbons. Frank Lloyd Wright understood what was happening. The same year Gibbons built his house, Wright wrote, "I believe that Romance—this quality of the *heart*, the essential joy that we have in living—by human imagination of the right sort can be brought to life again. . . . Our architecture . . . [will] become a poor, flat-faced thing of steel bones, box-outlines, gas-pipe and hand-rail fittings . . . without this essential *heart* beating in it. Architecture, without it, could inspire nothing." Today's owners of the Gibbons house have brought these qualities so beautifully to their restoration that Wright would have approved.

VANNA VENTURI HOUSE

PHILADELPHIA, PENNSYLVANIA

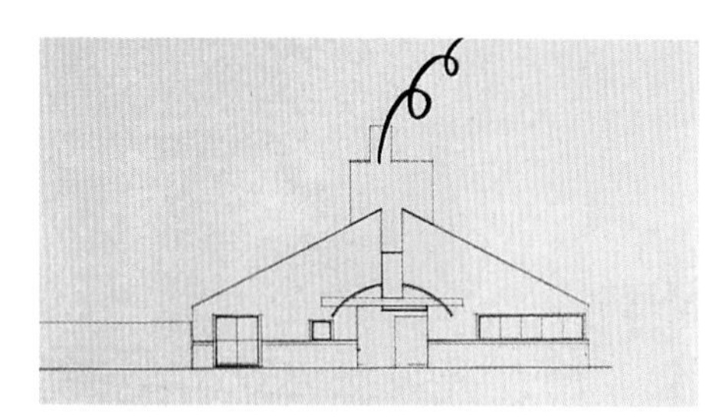

ABOVE: *Architect Robert Venturi describes the house he designed for his mother in 1964 as "an elemental house, like a child's drawing of a house." His own doodle illustrates the point.*

RIGHT: *In this famous photograph, Vanna Venturi sits in the loggia of the house, now recognized as one of the most important buildings of the late twentieth century.*

ASK ANY CHILD TO DRAW A PICTURE OF A HOUSE, AND WHAT usually appears looks very much like the doodles Robert Venturi drew of the house he built in 1964 for his widowed mother, Vanna, even to the curlicue of smoke puffing from the chimney. This wasn't just a bit of whimsy on the architect's part, though he admits to "using a joke to get to seriousness." In a look back at his mother's house twenty-five controversial years later, Venturi explained, "In its sheltering manner, with its gable roof, central door, ordinary windows and chimney, it looked like an elemental house, like a child's drawing of a house."

When you first approach the site, it's hard to see why a house that looks so plain made architects so angry at the time, and why it infuriated the neighbors. Today, architects realize, as one colleague put it, that "it's perhaps the most important building of the late 20th century . . . a revolutionary design that challenged and changed the course of architecture."

It had taken Venturi five years to perfect his design for his mother's house, and he knew he might be in for trouble. "It hurts when you are going against the grain. . . . I remember how hard this building was for me to arrive at," he revealed later. But there was no other way for a budding, brilliant young architect who had "walked on air" during his first trip to Europe, and who saw the Rome of both ancient *and* modern times as "architectural heaven." Venturi decided he would be guided "not by habit but by a conscious sense of the past, by precedent thoughtfully considered." To him, modern architects had thrown out the baby with the bathwater or, as he put it, "the variety with the vulgarity." "Less is a bore," Venturi said, in what may be the second-most-famous utterance any modern architect has ever made, the first being Mies van der Rohe's "Less is more."

Venturi's mother, the daughter and wife of Italian immigrants, imposed a few practical restraints: she didn't have a lot of money to spend, and she didn't want anything pretentious. She was a fantastic person, Venturi remembers, cultivated, a socialist and pacifist, whose "sound but unorthodox positions worked to prepare me to feel almost all right as an outsider." She loved what her son created for her, and she relished telling the architecture students who soon began making pilgrimages to the house, "This facade will tell you a lot of stories, if you will listen to it." They're still coming from all over the world. So are the historians, and architects, at last. Her stories about how it came to be are now architectural history.

Homeowners Tom and the late Agatha Hughes bought the house in 1975. Professor Hughes feels that the house is wonderfully American in its complexity and contradictions, and explains that his wife saw the house as "warm and embracing, not austere and intellectual as some believe."

After Vanna Venturi died in 1975, the house became the property of historian Thomas Hughes and his late wife, Agatha, an editor, ceramicist, and educator. Together they loved it just as much as she did, and they were still discovering fresh things even after several decades. "They have taken good care of it," Venturi says affectionately. He comes over to approve the color each time the stucco facade has to be repainted, and he agreed with Agatha's idea to paint it mauve someday! The Hugheses understanding and appreciation of the house became so ingrained they could not bear to leave it, despite the professional temptations that have come Professor Hughes's way as a distinguished pioneer historian of science and technology. Hughes relishes living in another pioneer's masterpiece, although it surely took Venturi longer than any medieval craftsman needed to convince his guild he qualified as a master.

For years, Venturi paid a price for violating the architectural tabus of the sixties, beginning with his mother's house, and later in works created in partnership with architect Denise Scott Brown, his wife. Commissions were sparse; their work was scorned or ignored. Painting the facade green was the least of their heresies, even though Venturi delights in pointing out it was good enough for the tsar's Winter Palace. But hadn't Marcel Breuer said a house could not be that color? Venturi lists other violations: the windows in his mother's house are real windows, "holes in the wall," he calls them, not the interruptions of wall, the absence of wall in the glass boxes of modern architecture. At the time, this idea was outrageous, "believe me," he insists. The lunette window upstairs was particularly daring! The inside of the house is enclosed space, a private place, with walls and corners, "not a glazed pavilion exposed to the elements and public view." A sloping roof was all right in those days if it was on a shed, but on a modern dwelling, of course, it had to be flat. On his mother's house, the two slopes of the roof meet on the long facade, not on the side as gables usually do, and form a huge pediment split down the middle. The arch above the front entrance, and those inside, are ornamental, not structural. So are the moldings on the facade and inside, and they are too high. Denise Scott Brown went so far as to compare the curved molding over the loggia, at least when Vanna Venturi was sitting out there, to a halo on a Renaissance painting of a Madonna. Venturi concludes, with irony, "You learn a lot from being perverse." His detractors finally did, too, and Venturi agrees it's hard to see now what the fuss was all about, now that the house has become an icon.

For Tom Hughes, the 1,800-square-foot house provides an environment that expresses values that are very congenial to him. Our generation grew up, Hughes

The rear of the house shows the lunette window in the second-floor bedroom that caused such a fuss at the time.

explains, with the values exemplified by modern mechanical technology: order, system, and control. He relishes today's very different, postmodern values: the interest in heterogeneity instead of rigid order, in control that's distributed, not hierarchical. This house is wonderfully American, Hughes believes, in its will to save traditional values and yet move along to new statements for a new culture. It is complex and full of contradictions, just as our diverse culture is full of contradictions. He agrees with Venturi's "Gentle Manifesto," which opened his seminal 1966 book on architectural theory, ***Complexity and Contradiction in Modern Architecture.*** "I am for messy vitality over obvious unity . . . " Venturi wrote, "[for] the difficult unity of inclusion rather than the easy unity of exclusion."

With her sophisticated artist's eye, Agatha Hughes cared for the house as the work of art Venturi finally realized he had created. But, she insisted, "it's not a solemn house. It likes to have fun." The half arc of the dining room ceiling, the diagonal and curved

Mrs. Hughes liked to describe the fireplace in the living room as "juggling for space" with the staircase and the front entrance behind it.

walls within the basic rectangle of the house, the whimsical "nowhere stair" with treads and risers reversed, the way the entrance, the living room fireplace, and the stairway to the upstairs study all juggle for space, illustrate her point.

Her witty touches in every room still enhance what Venturi calls "a little house with big scale." The entrance loggia is too large, the pediment too deep, the chimney too tall. Inside, the fireplace is too big and the mantel too high. But that explains the "mansion-sized" andirons Agatha placed in the oversized fireplace, the baroque candelabra, the tiny bentwood chair sitting next to a giant dictionary in the lunette window in the studio. When Denise Scott Brown called the house a "puppy with large feet," she was referring to Venturi's rich historical references in the house. Venturi often mentions his great love of Michelangelo's Porta Pia in Rome, the impact on his work of Vanbrugh's Blenheim Palace in England. Scott Brown adds, "If American houses are Queen Anne in front and Mary Anne behind, this one is Palladian in front and Aalto at the back—the 'nowhere stair' . . . is Furness with De Chirico out of Colonial America."

You might think Venturi and Denise Scott Brown would loathe places like Levittown—with its houses that could be put up thirty a day, and with its pastiche of historical styles that fulfills owners' fantasies of living in Tudor mansions or Tara. They don't. Venturi staked his reputation from the beginning on the position that architects have a lot to learn from places like Levittown—and Las Vegas. Most modern architects, they write, are "Experts with Ideals," interested in building "for Man rather than for people," i.e., "to suit their own particular upper-middle-class values, which they assign to everyone."

The half arc of the dining room ceiling is an example of the house "having fun," as Mrs. Hughes put it.

Venturi and Scott Brown like to see plastic flowers in the windows of the apartment house he designed for the elderly in a modest Philadelphia neighborhood. "Main Street is almost all right," they say. "Landscapes should include buildings plain and fancy." They relish the connections they find between the past and present, between, for example, the triumphal arches of ancient Rome and today's billboards along the Strip—between "the decorated shed" that was the Italian palazzo and that today is a high-rise Howard Johnson motel! "The Italian landscape has always harmonized the vulgar and the Vitruvian," they wrote in *Learning from Las Vegas.* "Naked children have never played in our [American] fountains, and I. M. Pei will never be happy on Route 66." Venturi and Scott Brown *are* happy there, and want us to be. "Sprawl and strip we can learn to do well," they insist, and while trying to show us the way, "they have changed the way we see the world."

THE HOUSE FOR THE NEXT MILLENNIUM

EXIT 2000, UNITED EXPRESSWAY 21

The House for the Next Millennium was a winner in "The Architect's Dream" competition. Entrants were asked to fulfill their fantasy, to create "what [they] could have if [they] could have anything."

THE POET EMILY DICKINSON HARDLY EVER LEFT HER FATHER'S austere brick house in the small college town of Amherst, Massachusetts. Yet in 1862, at midpoint in her reclusive existence, she wrote twelve lines that resonate perfectly with the cosmopolitan spirit of the two young American architects, Gisue Hariri and Mojgan Hariri, as they designed their House for the Next Millennium, well over a century later:

I dwell in Possibility,
A fairer House than Prose—
More numerous of Windows—
Superior—for Doors

of Chambers as the Cedars—
Impregnable of Eye—
And for an Everlasting Roof—
the Gambrels of the Sky—

Of visitors—the fairest—
For Occupation—This—
The spreading wide of my narrow Hands
To gather Paradise—

"Possibility" is a word you hear every day at Hariri & Hariri's crisp, compact white studio in Greenwich Village, just off Fifth Avenue. "Bold" . . . "visionary". . . "prophetic" are words heard frequently to describe these Iranian-born and Cornell-trained sisters. "They exemplify the spirit of the next generation . . . they're working in a space-age mode," say their admirers. "Gathering Paradise" isn't just a matter of words for the Hariris. Exquisite models of their completed works and projected designs fill their two-story atelier. "We're pragmatists," they say. "We like to build." And they do: dramatic lofts, apartments, and commercial studios in Manhattan; renovations and additions to contemporary and turn-of-the-century houses in the country; even a speculative project with a commercial developer in northern Virginia, which Gisue describes as "a testing ground for home buyers preparing for twenty-first-century living." "We like to experiment and think of the future," they admit. "We are trying to bring theory and practice together in one place."

The "NeXt House," their House for the Next Millennium, was a winner in "The Architect's Dream" competition for an exhibition at the Contemporary Arts Center in Cincinnati, Ohio, in 1993. "By dream," the exhibit curator said when he issued his invi-

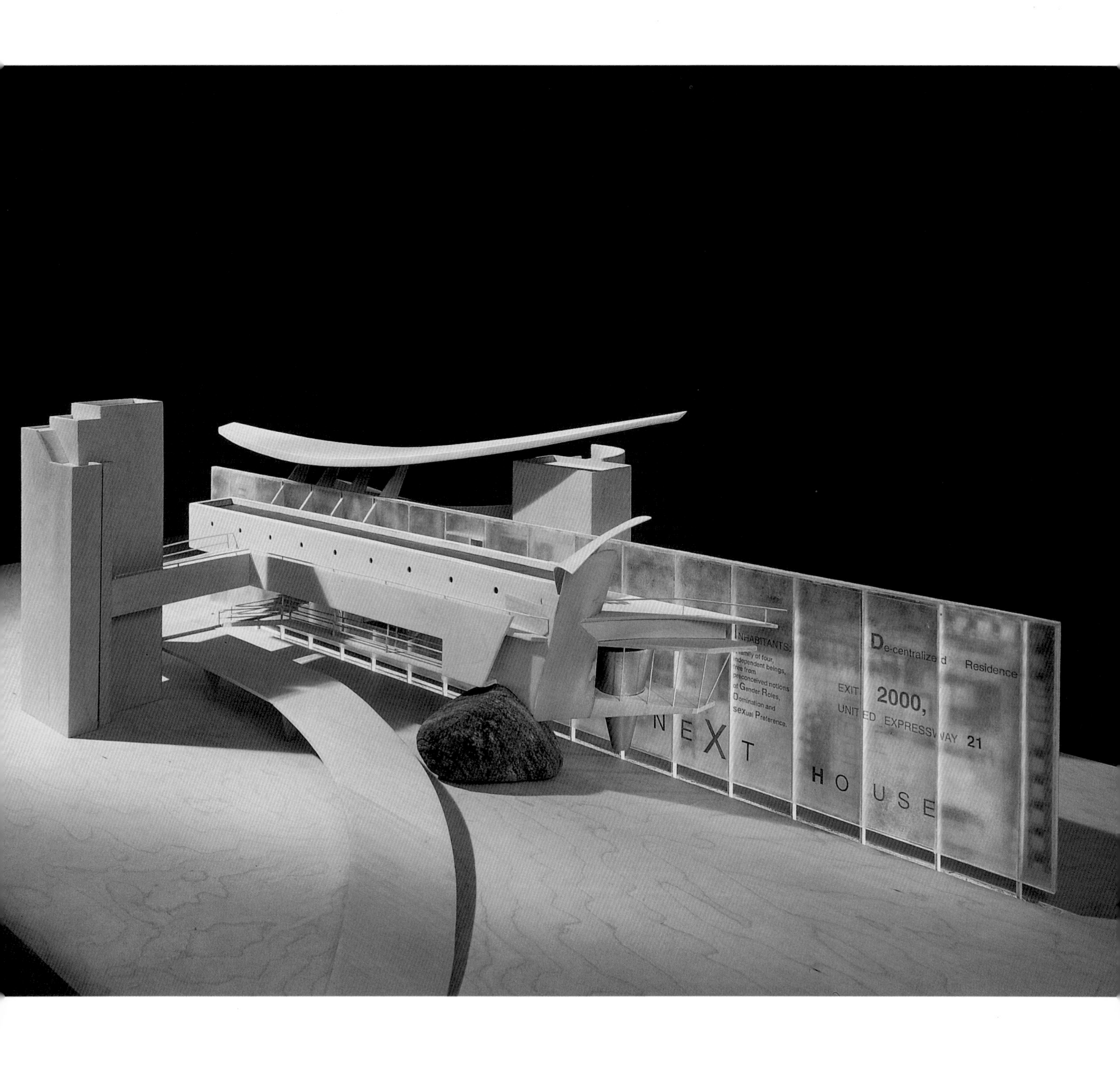
INHABITANTS:
family of four,
independent beings,
free from
preconceived notions
of Gender Roles,
Domination and
sexual Preference.
De-centralized Residence
EXIT 2000,
UNITED EXPRESSWAY 21
NEXT
HOUSE

"It's like being in a Mobius strip," Gisue Hariri says of this staircase, in a New York penthouse, made of a single sheet of burnished steel that curves in a logarithmic spiral.

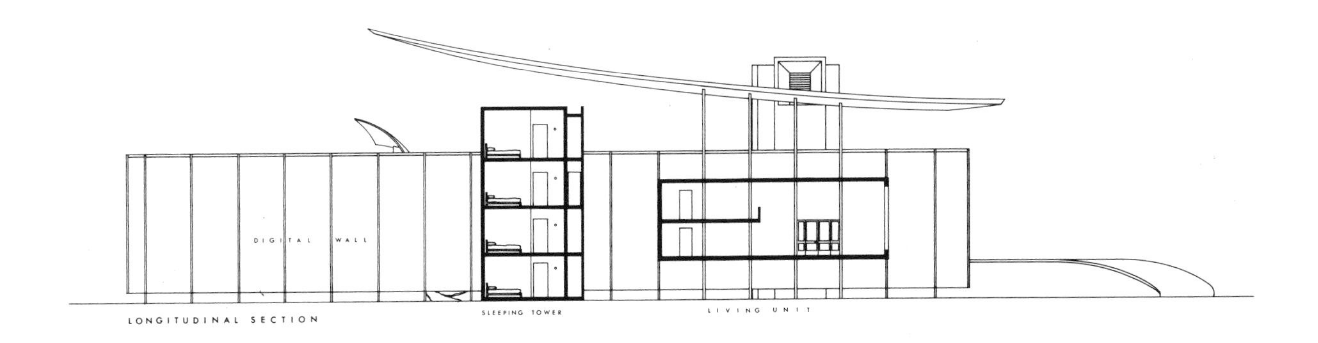

ABOVE: *A longitudinal section of the house shows the sleeping tower and its relation to the digital wall. Each bedroom will have a dream recording device—the images it captures during the night can be replayed and studied the next day.*

LEFT: *This preliminary sketch shows the swimming tube in their House for the Next Millennium.*

tations, "we imply wish fulfillment or fantasy, also perfection, what we could have if we could have *anything*." The Hariris' initial sketches, their doodles, didn't bear the slightest resemblance to a child's rendering of a house—nor to Venturi's elemental image of his mother's house. "We need a totally new outlook on living space for the next millennium," they insist. Their fantasy makes the gigantic technodwellings of today's cyberbarons look tame.

Hariri & Hariri are quite specific about some elements in their twenty-first-century house. The occupants? "A family of four independent beings free of preconceived

ABOVE: *Hariri & Hariri are working in a space-age mode, their admirers say. The House for the Next Milennium (viewed from the rear) is a perfect example.*

OPPOSITE: *In this realized 1992 project in Ontario, the window design echoes the horizontal markings on the bark of birch trees in nearby woods.*

notions of gender roles, domination, and sexual preference." Location? Off Exit 2000, United Expressway 21. Living quarters? No large, multifunctional rooms like the family room anymore, because they no longer serve any useful function. Instead, "the architecture of the main spaces will be reduced to minimum units, prefabricated and available off the shelf, potentially, to keep the house of tomorrow economically feasible."

So far, not totally far out. After all, front parlors like the Lincolns' disappeared years ago, and the profile of the typical American family undergoes constant reshaping. What makes Hariri & Hariri's design so extraordinary is the liquid wall that runs the entire length of the structure, a kind of transparent spine made out of LCD (liquid crystal display) panels, rather than a traditional facade. Habitats for sleeping, working, and entertainment, arranged vertically in tower form, are plugged "metaphorically, structurally, and technologically" into this digital wall, "flowing with information and programming, activated by touch and constantly changing." Through it, the occupants will communicate with each other, do their shopping, perform their jobs, entertain themselves, "find out what people are doing, what's cooking," Gisue explains. The house will also have a swimming tube for physical fitness, and a recycling device "for reducing, recycling and reusing." A network of passageways will connect the decentralized units of the house, and in Hariri & Hariri's scheme, these become critical features, because in them the inhabitants will "unplug" themselves for a while from the virtual world. "They are spaces for contemplation, for physical fitness and spiritual well-being. Here you will face the landscape, experience the sunset. You take the time to be actual as opposed to virtual," says Gisue, who lectures widely and teaches graduate-level design at Columbia University. She and Mojgan see larger possibilities for the digital wall. It could become a building block, like Frank Lloyd Wright's textile block, functioning architecturally as a glass partition, a window, or a wall inside or outside a house.

Their new building block may take another five or ten years to be realized, they are told by experts intrigued by their concept. LCD technology is still very expensive, and not yet as transparent as the Hariris would like. But they continue to investigate its potential for a twenty-first-century world that through digital technology, they believe, will be more interconnected, more open, and less focused on class or gender. In their own lives and work, the Hariris have already bridged two very different cultures from

Sisters Mojgan and Gisue Hariri, architects born in Iran and trained at Cornell, have their atelier in Manhattan.

the East and West. Their earlier work reflects Eastern architectural tradition. "The contrast of materials, the use of craft and hand-made things, the geometry and tactility are what we experienced in Iran," says Mojgan. They prefer heavy materials, "like thick, solid stone, and brick walls made the old way, with real plaster as opposed to wood-frame construction," which, to Mojgan, "feels flimsy and hollow." "We carry a lot of memories with us all the time," they say. For over a decade now, Hariri & Hariri have turned these memories into elegant, sensuous reality, beginning in 1986 with a magnificent staircase—partly straight and partly spiral—for a Soho loft. The balustrade is made of a "single sheet of burnished steel, curving in a logarithmic spiral . . . like being in a Mobius strip," Gisue explains. To be practical, she adds, the steel was treated not to show children's finger marks.

Others perceive an affinity in Hariri & Hariri's work with the Arts and Crafts movement in stunning hand-finished features such as a smooth desert red plaster wall, a rough stucco-textured mustard yellow cylinder housing a refrigerator, or a brushed-steel mantel. To relate one site to its natural surroundings, for example, Hariri & Hariri designed windows that reflect the horizontal markings on the bark of birch trees in nearby woods. Gisue reveals that she and Mojgan got the idea for the name of their firm when they visited Greene & Greene's Arts and Crafts masterpiece, the Gamble House.

It's understandable that Hariri & Hariri, Iranian-born, would be acutely aware of the chaos and conflict that come with change from one age to another. Some of their more fantastical designs deal with reconciliation and healing, building bridges between cultures, genders, and classes. The sisters deny that they are building "housing for the schizophrenic homeless in cyberspace." On the contrary, Gisue argues, "Now that we are in the post Industrial Revolution, houses are important again."

Even so, their House for the Next Millennium is a quantum leap from Orson Fowler's pre–Civil War octagon design, Buckminster Fuller's 1940s Dymaxion House, Frank Lloyd Wright's Usonian houses, and the many radical concepts from yesterday's tomorrows—from the houses of the future made of concrete, plastic, aluminum, plywood, glass; the house that could be cleaned by turning the hose on it, or shipped to a new location rolled up in a tube; the house where "a busy woman can rush in the front door, or in from the kitchen, glance at the clock and with one switch, light her cigarette and turn on the radio."

Architect Bart Prince has been called "the Jules Verne of modern architecture." His design for a New Mexico residence epitomizes the maverick spirit and daring displayed over the last century in American home design and construction.

A major danger, the Hariris fear, is that people could lose human contact in tomorrow's digital universe. Already there may be too much focus on the interior self in these technohouses, critics say. The Hariris recognize the paradox, "the human desire to be connected and disconnected from city life at the same time." They know that habitats of tomorrow must recognize there is a human soul that needs to be nourished. On the other hand, if Emily Dickinson could see so many possibilities from her isolation, perhaps their fear is unfounded.

NOTES

How If This House Could Talk Came to Be

viii *"The strength of a nation":* Quoted in Barbara J. Howe et al., *Houses and Homes: Exploring Their History* (Nashville: American Association for State and Local History, 1987), 1.

ix *But all he did:* Robert Frost, "America Is Hard to See," in *Collected Poems, Prose, and Plays* (New York: Library of America, 1995), 431.

I: Living in Art

Tlingit Clan House, Wrangell, Alaska

6 *"If this house could talk":* Author interview with Marge Byrd, Wrangell, Alaska, December 15, 1994.

6 *"A prime requisite":* Edward L. Keithan, *Monuments in Cedar* (Ketchikan, Alaska: Roy Anderson, 1945), 67.

6 *"as if a tree had fallen":* John Muir, *Travels in Alaska* (1915; reprint, San Francisco: Sierra Club Books, 1988), 62.

6 *The magnitude of the ruins:* Ibid., 59.

7 *with the figure:* Ibid., 60–61.

Frank Lloyd Wright's Storer House and Auldbrass Plantation, Hollywood, California / Yemassee, South Carolina

10 *"muv, I love you":* Quoted in Robert L. Sweeney, *Wright in Hollywood: Visions of a New Architecture* (Cambridge, Mass. and London: MIT Press, 1994), 63.

10 *"a love of the beautiful":* In David Larkin and Bruce Brooks Pfeiffer, eds., *Frank Lloyd Wright: The Masterworks* (New York: Rizzoli International Publications and the Frank Lloyd Wright Foundation, 1993), 228.

10 *"He tossed off":* Joseph Giovannini, "Architecture's Genius, America's Hero" in a publication for the Chrysler and *House Beautiful* Wright exhibition, 1996–1997.

12 *"We would take":* In Larkin and Pfeiffer, op. cit., 135.

13 *"I believe a house":* Ibid., 239.

13 *"It looked like":* In Bruce Brooks Pfeiffer, *Frank Lloyd Wright: His Living Voice* (Fresno, Calif.: The Press at California State University, Fresno, 1987), 42.

13 *"When I come home":* Charlie Rose interview with Joel Silver, February 9, 1994.

13 *"This is remarkable":* Ibid.

13 *"Color... would help a lot":* In Sweeney, op. cit., 63.

13 *"If you want good design":* In Pfeiffer, op. cit., 137.

15 *"the most friendly":* In Larkin and Pfeiffer, op. cit., 197.

15 *"to get the outside inside":* Ibid., 268.

15 *"It's insane":* Author interview with Joel Silver, Auldbrass Plantation, November 2, 1995.

16 *"no other piece":* Author interview with Eric Lloyd Wright, August 1, 1995.

16 *"Hope dies!":* Ibid.

17 *"The hexagon":* In Pfeiffer, op. cit., 184.

19 *"A house should have repose":* In Larkin and Pfeiffer, op. cit., 149.

Greene & Greene's Gamble House, Pasadena, California

20 *"to make the whole":* Quoted in Edward R. Bosley, *Gamble House: Greene and Greene* (London: Phaedon Press: 1992).

23 *All art-loving people:* In *Jeanette A. Thomas, Images of the Gentle Horse: Masterwork of Greene & Greene* (The Gamble House, University of California, 1989), 25.

23 *"there should be":* In Larkin and Pfeiffer, op. cit., 132.

23 *"Have nothing in your houses":* In Guy Wilson, "Divine Excellence: The Arts and Crafts Life in California," in *The Arts and Crafts Movement in California: Living the Good Life* (New York: Oakland Museum and Abbeville Press, 1993), 17.

24–25 *"There was a reason":* In Randell Makinson, *Prairie School Review* 5, no. 4 (1968): 11.

25 *"Anyone can build":* In Larkin and Pfeiffer, eds., op. cit., 161.

The Robert Frost Farm, Derry, New Hampshire

26 *"R. Frost has moved":* Quoted in Lesley Frost: *New Hampshire's Child: The Derry Journals of Lesley Frost* (Albany:

State University of New York Press, 1969), 2.

26 *"The core of all my writing":* Robert Frost to Robert Chase, March 4, 1952.

28 *"Our farm has interesting places":* Lesley Frost, *New Hampshire's Child,* bk. 5 (1907–1908), 78.

28 *"playing school":* Ibid., Book 3 (1905–1907), 50.

28 *"organized subjects":* Lesley Frost, "Robert Frost Remembered," *The American Way Magazine* 7, no. 2 (March 1974), 15.

28 *"before we caught on":* Lesley Frost, "Our Family Christmas," *Redbook Magazine,* December 1963, 98.

28 *"rumple [students'] brains":* In Lesley Lee Francis, *The Frost Family's Adventure in Poetry: Sheer Morning Gladness at the Brim* (Columbia and London: University of Missouri Press, 1994), 2.

28 *"Think!":* In Arnold Grade, ed., *Family Letters of Robert and Elinor Frost* (Albany: State University of New York Press, 1972), 52.

28 *"The rule at the Frost house":* Derry Farm video.

28 *"Reading was most important":* Lesley Frost, "Robert Frost Remembered," 15.

28 *"There was poetry, poetry and more poetry":* Lesley Frost, *New Hampshire's Child,* introduction, pp. 7–8.

29 *"Our hearts":* Ibid., pp. 7–8.

29 *"I learned that":* Ibid., 3.

29 *"coincide with specific topics":* Francis, op. cit., 31.

29 *"We didn't know":* Lesley Frost, *New Hampshire's Child,* 5, 9.

30 *"a way of taking life":* In William H. Pritchard, *Frost: A Literary Life Reconsidered* (New York and Oxford: Oxford University Press, 1984), 58.

30 *"inspiration doesn't live":* In Edward Connery Lathem and Lawrance Thompson, eds., *Robert Frost: Poetry and Prose* (New York: Holt, Rinehart & Winston, 1972), 291.

30 *"chore-time–playtime":* Lesley Frost, *Derry Down Derry–A Narrative Reading of Poems of Robert Frost* (Album #FL 9733, Folkways and Services Corp., 1961).

30 *"poetry has to heighten":* In Lathem and Thompson, op. cit., 315–316.

30 *"It's our business":* Ibid., 253.

30 *"It is not fair":* Ibid., 270.

31 *"It will be a relief":* In Grade, op. cit., 111.

31 *"they came right up out of the ground":* Lesley Frost, *New Hampshire's Child,* bk. 4 (1906–1907), 60.

31 *"who was always accessible":* Francis, op. cit., 4.

31 *"castle, province":* Lesley Frost, *New Hampshire's Child,* introduction, 6.

31 *"Catch her getting":* In Lathem and Thompson, op. cit., 271.

31 *"very old and worn":* Lesley Frost, *New Hampshire's Child,* bk. 5 (1907–1908), 59.

31 *"More than once a day":* Ibid., 66.

32 *"Mama won't like it":* Ibid., bk. 6 (1908–1909), 27; bk. 3 (1905–1907), 108.

32 *"Papa generally":* Ibid., bk. 3 (1905–1907), 108.

32 *"My, my, what sorrow":* In Grade, op. cit., 209–210.

32 *"The power she exercised":* Lesley Frost, *New Hampshire's Child,* p. 6.

32 *"Together wing to wing":* Robert Frost, "The Master Speed," op. cit., 273.

32 *"there is at least":* In Lathem and Thompson, op. cit., 344.

33 *He brought:* In Lawrance Thompson and R. H. Winnick, *Robert Frost,* 1-vol. ed. (New York: Holt, Rinehart & Winston, 1981), 514–515.

33 *Well away:* Robert Frost, "On the Sale of My Farm," op. cit., 519.

II: George Washington Didn't Sleep Here

The Adams Family's Old House, Quincy, Massachusetts

38 *"residing at the Court":* Quoted in National Park Service, Department of Interior, Adams National Historic Site brochure.

38 *"Tis Domestick happiness":* Abigail Adams's diary on board ship, May 1788, Adams House Archives.

38 *"You cannot crowd":* In Edith B. Gelles, *Portia: The World of Abigail Adams* (Bloomington and Indianapolis: Indiana University Press, 1992), 122.

38 *"a very Genteel Dwelling House":* In Adams National Historic Site brochure.

40 *"most sadly disappointed":* In *The Adamses at Home: Accounts by Visitors to the Old House in Quincy, 1788–1886*

(preprinted from the *Publications of the Colonial Society of Massachusetts*, vol. 45), p. 10.

40 *"I found my estate":* Ibid., 11.

40 *"beneath himself":* In John Ferling: *John Adams: A Life* (Knoxville: University of Tennessee Press, 1992), 298.

40 *"an ordeal path": Adams Quotations* (a compilation by Adams Historic Site), 38.

40 *"enjoy the cool Evening": Adams Quotations,* p. 31.

40 *"sweet little farm":* In Paul Nagel, *Descent from Glory: Four Generations of the John Adams Family* (Oxford: Oxford University Press, 1983), 48.

40 *the most insignificant:* In Adams, *Quotations,* 60.

41 *"ten talents":* In Bruce Miroff, *Icons of Democracy: American Leaders as Heroes, Aristocrats, Dissenters, & Democrats* (New York: Basic Books, 1993), 67–68.

41 *"not the Tincture of Ha'ture":* In Gelles, op. cit., 125.

41 *"there is danger from all men":* John Adams statue, Quincy, Mass.

41 *"When and where": Adams Quotations,* 37.

41 *"It would seem":* In Joseph J. Ellis, *Passionate Sage: The Character and Legacy of John Adams* (New York and London: W. W. Norton & Co., 1993), 123.

41 *"Rage a little":* Ibid., 176.

41 *"He lets me do":* In Wilhelmina S. Harris, *Adams National Historic Site, A Family's Legacy to America* (Washington, D.C.: U. S. Department of the Interior, National Park Service, 1983), 5.

43 *"Every Scrap":* In L. H. Butterfield, *The Papers of the Adams Family: Some Account of their History, Proceedings of the Massachusetts Historical Society* 71 (1959), 335.

43 *My house:* Abigail Adams to Thomas Boylston Adams, December 25, 1800, Adams Historic Site files.

43 *"showed plainly enough":* Henry Adams, *The Education of Henry Adams* (New York: The Modern Library, 1931), 10.

43 *"Every room is full of history":* In *The Adamses at Home,* op. cit., 53–54.

43 *"abode of enchantment":* Ibid., 41.

43 *"a retreat":* In Butterfield, op. cit., 339.

43 *"Hot water":* In Nagel, op. cit., 161.

44 *by the Name of Peace field: Adams Quotations,* 35.

44 *"Montezillo":* In Ellis, op. cit., 58.

44 *"Our desires are moderate":* In Ferling, op. cit., 409.

44 *"Everything the best":* In *The Adamses at Home,* 25.

44 *"It need not be said":* Ibid., 15.

44 *"his manner of life":* Ibid., 22.

44 *"A simpler manner":* Henry Adams, op. cit., 10.

44 *"It is the great and foul stain": Adams Quotations,* 56.

44 *"You will find your father":* In Jack Shepherd, *The Adams Chronicles: Four Generations of Greatness* (Boston and Toronto: Little, Brown & Co., 1975), 220.

45 *"If we mean to have": Adams Quotations,* 62.

45 *"natural genius":* In Nagel, op. cit., 202.

45 *"I had to Act": Adams Quotations,* 32.

45 *"the best, dearest":* In Nagel, op. cit., 23.

45 *"ticklish temper":* In Shepherd, op. cit., 196.

45 *"grotesque rhetoric" and "obnoxious principles":* In Ferling, op. cit., 412, 309.

45 *"so amiable": Adams Quotations,* 7.

45 *"descending so smoothly":* In Nagel, op. cit., 129.

45 *"had not the smallest chip":* In *The Adamses at Home,* 14–5.

45 *Do no wrong: Adams Quotations,* 33.

46 *I am a man:* Ibid., 7–8

46 *"by the ease":* In *The Adamses at Home,* 35.

46 *"Political movement": Adams Quotations,* 38.

46 *"Where else":* In Frederick S. Voss, *The Smithsonian Treasury: The Presidents* (Washington D.C.: Smithsonian Institution, 1991), 20.

46 *"Had I stepped":* In L. H. Butterfield, "Tending a Dragon-killer: Notes for the Biographer of Mrs. John Quincy Adams," *Proceedings of the American Philosophical Society* 118, no. 2 (April 1974), 169.

47 *"It was so cold":* "Mrs. John Quincy Adams' Narrative of a Journey from St. Petersburg to Paris in February 1815," *Scribner's Magazine* 34, no. 4 (October 1903), 448–463.

47 *Among all the great characters:* In Butterfield, op. cit., 169n.

47 *To the boy:* Henry Adams, op. cit., 17–18.

47 *"My lot in marriage":* In Shepherd, op. cit., 217.

47 *"I can neither live":* In Nagel, op. cit., 101.

47 *"I wish I could":* In *Adams National Historic Site* biography of Abigail Adams.

50 *He likes to have:* In *The Adamses at Home*, 31.

50 *all about me:* In Ellis, op. cit., 189.

50 *when the door:* Henry Adams, op. cit., 13.

50 *"too much of my time":* In Nagel, op. cit., 33.

51 *Train them to Virtue:* In Gelles, op. cit., 143.

51 *"The family cared little":* In Malcolm Freiberg, "From Family to Nation: The Old House Becomes a National Historic Site," *Proceedings of the Massachusetts Historical Society* 98, (1986) 74.

51 *"commerce, luxury": Adams Quotations*, 59.

James Madison's Montpelier, Orange, Virginia

52 *"My beloved":* Quoted in Conover Hunt-Jones, *Dolley and the Great Little Madison* (Washington, D.C.: American Institute of Architects Foundation, 1977), 15.

52 *"Adam and Eve in their Bower":* In Ralph Ketcham, *James Madison: A Biography* (Charlottesville: University Press of Virginia, 1990), 621.

52 *"the best farmer in America":* Ibid.

52 *"too coarse and dry":* In Merrill Peterson, *James Madison: A Biography in His Own Words* (New York: Newsweek, 1974), 28.

54 *The principles and Modes of government:* Ibid., 26.

54 *"a necessary misfortune":* In Ketcham, op. cit., 298.

54 *"if people have enough virtue":* Dr. Josephine F. Pacheco, In "Launching of a Landmark," National Trust for Historic Preservation booklet (Washington, D. C.: 1997), 7.

54 *"mutual animosities":* James Madison, *Federalist Papers*, no. 10.

54 *"If any American":* James MacGregor Burns, *The Vineyard of Liberty* (New York: Alfred A. Knopf, 1982), 27.

54 *"Two brothers":* Paul Jennings, *A Colored Man's Reminiscences of James Madison* (Brooklyn: George C. Beadle, 1865), 3, 17.

54 *"the greatest man":* In James Monroe Smith, ed., *The Republic of Letters: The Correspondence between Thomas Jefferson and James Madison, 1776–1826,* vol. 1 (New York and London: W. W. Norton & Co.), 38.

55 *"putting up and pulling down":* In Hunt-Jones, op. cit., 61.

55 *"They always made you glad":* In Hunt-Jones, op. cit., 74.

55 *"no heavier":* Author conversation with Rev. James Dunne.

55 *"the House . . . plain but grand":* In Ketcham, op. cit., 427.

57 *"from the dismantled palace":* In Ketcham, op. cit., 614.

57 *"Everything displayed":* In Hunt-Jones, op. cit., 89.

57 *"Mrs. M.":* Harriet Martineau, *Retrospect of Western Travel*, vol.1 (New York: Haskell House Publishers, 1969), 193.

57 *"a bright story":* In Ketcham, op. cit., 511.

58 *"frankness and ease":* In Smith, op. cit., vol. 3, 1563.

58 *"His house would be":* In Ketcham, op. cit., 607.

59 *"a country schoolmaster":* In Hunt-Jones, op. cit., 11.

59 *"the best informed":* In Ketcham, op. cit., 201, 73, 112, 201.

59 *Mr. Madison:* Ibid., 476.

59 *"better informed":* Drew R. McCoy, *The Last of the Fathers: James Madison and the Republican Legacy* (Cambridge: Cambridge University Press, 1989), 27–33 passim.

59 *"a greater and far more":* In Ketcham, op. cit., 670.

59 *"Notwithstanding a thousand":* In McCoy, op. cit., 16–17.

60 *"She had no time for it":* Jennings, op. cit., 14–15.

60 *He was in a chair:* Martineau, op. cit., 190, 198.

61 *"To myself":* In James Monroe Smith, op. cit., vol. 3, 1967.

Abraham Lincoln's House, Springfield, Illinois

62 *"It isn't the best thing":* Quoted in Ruth Painter Randall, *Mary Lincoln: Biography of a Marriage* (Boston: Little, Brown & Co., 1953), 137.

62 *"He could split hairs":* In Cullom Davis, *Abraham Lincoln and the Golden Age of American Law,* Historical Bulletin no. 48 (Racine, Wis.: Lincoln Fellowship of Wisconsin, 1994), 12.

64 *"Mary insists":* In Michael Burlingame, *The Inner World of Abraham Lincoln* (Urbana and Chicago: University of Illinois Press, 1994), 248. Note: A sucker is a fish found in Illinois—the natives of the state used to be called that.

64 *"the man who is of neither party":* In David Herbert Donald, *Lincoln* (New York: Simon & Schuster, 1995), 189.

64 *"He rose grandly":* In Ruth Painter Randall, *The Courtship of Mr. Lincoln* (Boston: Little, Brown & Co.), 95.

64 *"without Mary Todd Lincoln":* In Burlingame, op. cit., 325.

64 *It is like the residence: Historic Furnishings Report,* Lincoln Home National Historic Site, 37.

65 *"of opposite natures":* In Justin Turner and Linda Turner, *Mary Todd Lincoln: Her Life and Letters* (New York: Alfred A. Knopf, 1972), 200.

65 *"was charmed":* In Randall, *Courtship,* 64.

65 *"creature of excitement":* In Randall, *Mary Lincoln,* op. cit., 3.

65 *"witty and fearless":* William Herndon, *Herndon's Life of Lincoln* (Greenwich, Conn.: Fawcett Publications, 1961), 344.

65 *"she could make":* In Turner and Turner, op. cit., 11 (n. 8).

65 *"not pretty":* In Randall, *Mary Lincoln,* op. cit., 91.

65 *"the people are perhaps":* In Donald, op. cit., 108.

65 *"I would rather marry":* In Randall, *Mary Lincoln,* op. cit., 14.

66 *"no children to ruin things":* Ibid., 138.

66 *"Children have first place":* In Burlingame, op. cit., 57.

66 *"It is my pleasure":* In Donald, op. cit., 109.

66 *"It is a great piece":* Ibid., 19.

67 *"the President's place":* In Turner and Turner, op. cit., 716.

67 *"owing to an* unlucky*": Historic Furnishings Report,* 22.

67 *About six o'clock:* Ibid, 42–43.

68 *"There is the animal":* Philip B. Kunhardt, Jr., Philip B. Kunhardt III, and Peter W. Kunhardt, *Lincoln, An Illustrated Biography* (New York: Alfred A. Knopf, 1992), 119.

68 *"the long and short":* In Randall, *Mary Lincoln,* 179.

69 *"Obscene clown":* Dr. Thomas Keiser, "Lincoln the Abused President," Third Annual Lincoln Colloquium, October 15, 1988 (Lincoln Home National Site, Sangamon County Historical Society and Lincoln Group of Illinois), 8.

69 *"the greatest man":* In Donald, op. cit., 239.

69 *The boss:* In Burlingame, op. cit., 13.

69 *"either in the garret":* In Michael Burlingame, *An Oral History of Abraham Lincoln: John G. Nicolay's Interviews and Essays* (Carbondale and Edwardsville: Southern Illinois University Press, 1996), 1.

69 *"Her hand is not soft":* In Randall, *Mary Lincoln,* op. cit., 84.

69 *"a matter of profound wonder":* In Donald, op. cit., 94.

69 *"was* not *a demonstrative man":* In Charles B. Strozier, *Lincoln's Quest for Union: Public and Private Meanings* (Urbana and Chicago: University of Illinois Press, 1987), 78.

69 *"he was a terribly firm man":* In Randall, *Mary Lincoln,* op. cit., 224.

69 *"Bring on the cinders":* Ibid., 78.

69 *"deep amiable nature":* In Turner and Turner, op. cit., 566, 284, 534.

69 *"My wife is as handsome":* Ibid., 114.

70 *"We must* both*":* In Donald, op. cit., 593.

70 *"Occupying the same rooms":* Beverly B. Buehrer, "Lincoln's Springfield Home: An Intimate View of a President's Life," *Early American Life,* February 1989, 39.

70 *"dimly lighted . . . room":* Noyes Miner, "Mrs. Abraham Lincoln: A Vindication" (Xerox of handwritten

paper from Illinois State Historical Library), 6.

71 *"restored to reason":* In Mark E. Neely, Jr., and R. Gerald McMurty, *The Insanity File* (Carbondale and Edwardsville: Southern Illinois University Press, 1986), 102.

71 *"what Beethoven was":* In Burlingame, *Inner World*, op. cit., 13.

71 *"I don't think":* Strozier, op. cit., 79.

71 *"a devoted wife":* Miner, op. cit., 11.

71 *"lofty pine trees":* In Turner and Turner, op. cit., xix.

III: The Truth About Tara

Rosedown Plantation, St. Francisville, Louisiana

74 *"commenced hauling":* Quoted in *Reflections of Rosedown*, 16 (booklet available from Rosedown Plantation, 12501 Highway 10, St. Francisville, LA 70775).

74 *"Saw Daniel Turnbull in town":* in typed draft ms for *Reflections.*

74 *"Cotton is on the levees":* typed ms for *Reflections.*

76 *"The influence":* John Hope Franklin and Alfred A. Moss, Jr., *From Slavery to Freedom: A History of Negro Americans*, 6th ed. (McGraw-Hill, Inc., 1988), 113.

76 *"the deluge of spoony fancy pictures":* In Laura Wood Roper, *FLO: A Biography of Frederick Law Olmsted* (Baltimore and London: Johns Hopkins University Press, 1973), 84.

76 *as rather more devoid:* Frances Anne Kemble, *Journal of a Residence on a Georgian Plantation in 1838–1839* (Athens: Brown Thrasher Books / University of Georgia Press, 1984), 63, 139.

77 *"the workmanship":* In *Reflections*, 16.

79 *We had 30 people:* Ibid., 19.

80 *"I've had so much company":* Martha Turnbull, *The Sixty Year Garden Diary of Martha Turnbull, Mistress of Rosedown Plantation, 1836–1896* (Rosedown Plantation publication, 1996), first entry, 1853.

80 *"I had to stint":* Ibid. (March 1, 1856).

80 *"sharp blue eyes":* Margaret Mitchell, *Gone With the Wind* (New York: Warner Books, 1993), 54.

80 *"The climate is a wretched":* typed draft ms of *Reflections.*

80 *"We were all getting sick":* Rosedown Archives, Turnbull-Brown-Lyman Papers 1:3, October 7, 1860.

80 *"Very dry and parching":* Turnbull, *Garden Diary*, June 18, 1838.

80 *"I find":* In Sally G. McMillen, *Southern Women: Black and White in the Old South* (Arlington Heights, Ill.: Harlan Davidson, 1992), 1.

81 *No matter how large:* Anne Firor Scott, *The Southern Lady from Pedestal to Politics, 1830–1930* (Charlottesville and London: University Press of Virginia, 1995), 31.

81 *"the saddest slavery":* In Harriet Martineau, op. cit., 192.

81 *"For women":* Scott, op. cit., 46.

81 *"akin to a psychological battleground":* Deborah Gray White, *Ar'n't I a Woman?: Female Slaves in the Plantation South* (New York and London: W. W. Norton & Co., 1985), 16.

81 *"I nebbah knowed whut it wah t'rest":* In Ira Berlin, Marc Favreau, and Steven F. Miller, eds. *Remembering Slavery* (New York: New Press, 1998), 273.

82 *"most [Southern women]":* McMillen, op. cit., 1, 99.

82 *"I will not make sacrifices":* Sarah Turnbull, letter date unclear, in Rosedown Archives.

82 *"300 hogsheads of sugar":* In *Reflections*, 31.

82 *"I only know":* Sarah T. Bowman letter, 1902, in Rosedown Archives.

82 *"Since the Federals landed":* In *Reflections*, 30.

83 *to ples to Cross the Mississippia:* In Ira Berlin, et al., eds., *Free at Last: A Documentary History of Slavery, Freedom, and the Civil War* (New York: New Press, 1992), 496.

83 *"Augustus said":* In *Reflections*, 30.

The Underground Railroad Houses, Ripley, Ohio

86 *"How near we may be":* Quoted in Andrew Ritchie, *The Soldier, the Battle and the Victory: Being a Brief Account of the Work of Rev. John Rankin in the Anti-Slavery Cause* (Cincinnati: Western Tract & Book Society, [1868?]), 111.

86 *"Everyone engaged in":* John P. Parker, *His Promised Land: The Autobiography of John P. Parker, Former Slave and Conductor on the Underground Railroad*, ed. Stuart Seely Sprague (New York and London: W. W. Norton & Co., 1996), 127.

86 *"I am now living":* Ibid., 86.

86 *"My house was in full view":* Ritchie, op. cit., 98.

88 *"fierce passions":* Parker, op. cit., 74.

90 *"Any other time":* Ibid., 94–95, 100.

90 *"easy of access":* Ibid., 146–147.

90 *"too late to get them away":* Ibid., 138–139.

90 *"I could see":* Ibid., 141.

91 *"must have gone off":* In Paul R. Grim, "The Reverend John Rankin, Early Abolitionist," typescript, n. d., n. p.

91 *"The real fortress":* Parker, op. cit., 86.

91 *"no man could love":* Grim, op. cit., n.p.

91 *"slavery's curse":* Parker, op. cit., 96.

91 *"The real injury":* Ibid., 25–26.

91 *"men lay around":* "Life of Reverend John Rankin Written by Himself in His Eightieth Year," typescript, 33.

91 *"I would remind":* In Edith M. Gaines, *Freedom Light* (Cleveland: New Day Press, 1991), 10.

92 *"All that my father":* reminiscences of Captain R. C. Rankin, for Henry Howe's *Historical Collections of Ohio,* n. d.

93 *"When her husband":* In Ritchie, op. cit., 107.

93 *Some were guided:* William Still, *The Underground Railroad: A Record of Facts, Authentic Narratives, Letter, &c.* (1872: reprint, New York: Arno Press and the New York Times, 1968), preface.

93 *"Life took on a new":* Parker, op. cit., 39.

93 *"My soul is vexed":* In Still, op. cit., 1872 edition, 65.

93 *"No day dawns":* In Charles Blockson, "Escape from Slavery: The Underground Railroad," *National Geographic,* July 1984, 13.

The Oaks of Booker T. Washington, Tuskegee, Alabama

94 *we were told:* Booker T. Washington, *Up from Slavery* (New York: Dover Publications, 1996), 10–11.

94 *"motley mixture":* Ibid., 13.

94 *"No race can prosper":* Ibid., 107.

94 *"no ordinary darkey":* In Louis Harlan, *Booker T. Washington: The Making of a Black Leader, 1856-1901* (London: Oxford University Press, 1972), 152 .

95 *"When at Tuskegee":* In Louis Harlan and Raymond W. Smock, *The Booker T. Washington Papers, vol. 5, 1899-1900* (Urbana: University of Illinois Press, 1976), 61.

96 *"I have found":* Washington, op. cit., 74.

97 *"nearly every colored church":* Eric Foner, *Reconstruction: America's Unfinished Revolution, 1863-1877* (New York: Harper & Row, 1988), 428.

97 *"[his] influence":* Franklin and Moss, *From Slavery to Freedom,* 250–251.

97 *"Benedict Arnold":* In Louis Harlan, "Booker T. Washington and the Politics of Accommodation," in *Black Leaders of the Twentieth Century,* ed. John Hope Franklin and August Meier (Urbana: University of Illinois Press, 1982), 6.

97 *"certainly the most distinguished":* W. E. B. Du Bois, *The Souls of Black Folk* (New York: Dover Publications, 1994), 26.

98 *"didn't believe":* Portia Washington Pittman, oral history interview, and in "The Oaks Furnishings Plan," n. p.

99 *"Her heart was set":* In Harlan, *Booker T. Washington,* 147.

99 *"She literally wore herself out":* Washington, op. cit., 96.

99 *"Steps not swept":* In Harlan, 273.

100 *"In many cases":* Washington, op. cit., 40.

100 *"The wisest among my race":* Ibid., 108–109.

101 *"went to their graves":* David Levering Lewis, *W. E. B. Du Bois: Biography of a Race* (New York, Henry Holt & Co., 1993), 258.

101 *"We must lay on the soul":* In Rayford W. Logan and Michael R. Winston, eds. *Dictionary of American Negro Biography* (New York and London: W. W. Norton & Co., 1983), 195.

101 *"a great man":* Martin Luther King, Jr., address, Atlanta, Georgia, August 11, 1967 (King Library and Archives), 10–12.

101 *"It cannot be said":* Glenn Loury, *One by One from the Inside Out: Essays and Reviews on Race and Responsibility in America* (New York and London: Free Press, 1995), 68.

IV: Forgotten Frontier

Wukoki Pueblo, Wupatki National Monument, Arizona

104 *"It is visible":* Quoted in Wukoki Pueblo National Park Service brochure, 1996.

106 *"You will go on long migrations":* In Harold Courlander, *The Fourth World of the Hopis* (New York: Crown Publishers, 1971), 32–33.

107 *"cosmic drama":* Frank Waters, *Book of the Hopi* (New York: Penguin Books, 1963), 27.

107 *"great sweep of human":* Courlander, op. cit., 13.

108 *"We Hopis":* In Jake Page, "Inside the Sacred Hopi Homeland," *National Geographic,* November 1982, 626.

109 *"My foundation":* Dan Namingha, in *Colores,* KNME-TV documentary.

109 *Cloud Brothers:* In Ramson Lomatewama, *Drifting Through Ancestor Dreams: New and Selected Poems* (Flagstaff, Ariz.: Northland Publishing Co., 1993), 50.

Cherokee Chief Vann's House, Chatsworth, Georgia

110 *"a friendly commerce":* Quoted in Philip Weeks, *Farewell, My Nation: The American Indian and the United States* (Arlington Heights, Ill.: Harlan Davidson, 1990), 12.

110 *We shall with great pleasure:* Ibid.

112 *Thus ended the life:* In Whitfield-Murray Historical Society, *Murray County's Indian Heritage,* 6.

112 *"[Vann] with all his errors":* In Thurman Wilkins, *Cherokee Tragedy: The Story of the Ridge Family and of the Decimation of a People* (New York: Macmillan Co., 1970), 34.

112 *"The whites are gradually":* "Moravian Diaries," vol. 1 (1805), 45, unpublished manuscripts translated by Carl Mavelshagan, Vann House files.

113 *It was right:* In Weeks, op. cit., 19.

114 *"[They] did not need":* William G. McLoughlin, "Who Civilized the Cherokees?" *Journal of Cherokee Studies,* (1988), 73.

114 *"I am sensible":* In Wilkins, op. cit., 113.

114 *Will anyone believe:* Ibid., 135–136.

114 *"[They were] more interested":* In *Murray County's Indian Heritage,* 26.

114 *"he knew": Vann House Furnishings Plan,* 4, n. 5.

114 *You asked us:* In Fergus M. Bordewich, *Killing the White Man's Indian: Reinventing Americans at the End of the Twentieth Century* (Garden City, N. Y.: Doubleday & Co., 1996), 46.

115 *"Until now":* "Moravian Diaries," vol. 2 (1829), 8.

115 *"the Indians were no more":* In Wilkins, op. cit., 210.

116 *"John Marshall":* Ibid., 229.

116 *"If I had known":* Weeks, op. cit., 25.

116 *Colonel Bishop [is] forted:* In *Murray County's Indian Heritage,* 52.

117 *"might have created":* Bordewich, op. cit., 58.

General Vallejo's Petaluma Adobe, Petaluma, California

118 *"It was a case" and "The great valley":* In Adair Heig, *History of Petaluma, A California River Town* (Petaluma, Calif.: Scottwall Associates, 1982), 5.

118 *"We were the pioneers":* In Madie Brown Emparan, *The Vallejos of California* (San Francisco: Gleeson Library Associates, University of San Francisco, 1968), 8.

120 *"My biography":* Ibid., 122.

120 *The house was of immense:* In James B. Alexander, *Sonoma Valley Legacy* (Sonoma: Sonoma Valley Historical Society, 1986), 68.

122 9. *"Voluptuous":* In Emparan, op. cit., 206.

122 *"[she] possess[ed]":* In Alan Rosenus, *General M. G. Vallejo and the Advent of the Americans* (Albuquerque: University of New Mexico Press, 1995), 159.

123 *On our arrival:* In Emparan, op. cit., 46.

123 *The General never met me:* Ibid., 46.

124 *You are the kind:* Ibid., 51–52.

124 *"We belong":* Ibid., 22.

124 *"A little before dawn":* Ibid., 26.

124 *"We have not been killed":* Ibid., 36.

125 *My heart grieved:* In Rosenus, op. cit., 138.

125 *"door every half hour":* In Emparan, op. cit., 206.

125 *"I left Sacramento half dead":* Ibid., 43.

125 *"the American ideas":* Ibid., 181.

125 *"In the hands of":* In Kevin Starr, *Americans and the California Dream* (New York and Oxford: Oxford University Press, 1973), 41.

125 *"a proud, indolent":* In Leonard Pitt, *The Decline of the Californios: A*

Social History of the Spanish-Speaking Californians, 1846-1890 (Berkeley and Los Angeles: University of California Press, 1971), 16.

126 *"the Anglo-Saxon's preoccupation":* Pitt, op. cit., 13.

126 *The Yankees are a wonderful:* Ibid., 240.

126 *"seeks his own good fortune":* Ibid., 278.

126 *"swollen torrent of shysters":* In Rosenus, op. cit., 200.

126 *"most shameless":* In Emparan, op. cit., 53.

126 *"I brought this":* In Pitt, op. cit., 241.

126 *It is a sad memory:* In Emparan, op. cit., 150.

127 [*In my dreams*]: Ibid., 150–151.

127 *If I must be cast:* In Starr, op. cit., 31.

127 *"the Wheel of Fortune":* In Emparan, op. cit., 107.

127 *"you could not fly":* Ibid., 230.

127 *"If Vallejo was not":* In Rosenus, op. cit., 41.

Lorenzo Hubbell's House, Ganado, Arizona

128 *"Ganado was a place":* David Brugge interview (interpreter Roberta L. Tso) with Asdzaa Bekash, Hubbell Trading Post National Historic Site Archives, November 8, 1971, 8.

128 *"the most hospitable man":* Quoted in Frank McNitt, *The Indian Traders* (Norman: University of Oklahoma Press, 1962), 217.

128 *"I never charged":* Lorenzo Hubbell, "Fifty Years an Indian Trader," *AAA Magazine*, December 1930, 29.

130 *"a first class scoundrel":* Ibid.

130 *"a money-grabbing":* McNitt, op. cit., 69.

130 *"Neither the trader":* Edward T. Hall, *West of the Thirties: Discoveries Among the Navajo and Hopi* (New York: Doubleday & Co., 1994), 142.

130 *"Out here":* Hubbell, op. cit., 24.

131 *"Mr. Hubbell, you are":* Ibid., 27.

135 *"What changes":* Ibid., 51.

135 *"is the goodwill":* David M. Brugge, *Hubbell Trading Post National Historic Site* (Tucson, Ariz.: Southwest Parks and Monuments Association, 1993), 79.

V: A Woman's Place

Rebecca Nurse Homestead, Danvers, Massachusetts

All quotes in this essay unless otherwise noted are from: Richard B. Trask, *The Devil hath been raised: A Documentary History of the Salem Village Witchcraft Outbreak of March 1692* (Danvers, Mass.: Yeoman Press, 1997), 7ff.

138 *"The Evil Hand":* In Marion L. Starkey, *The Devil in Massachusetts: A Modern Enquiry into the Salem Witch Trials* (Garden City, N. Y.: Doubleday/Anchor Books, 1989), 42.

140 *"honored and dear mother":* In Janice Schuetz, *The Logic of Women on Trial: Case Studies of Popular American Trials* (Carbondale and Edwardsville: Southern Illinois University Press, 1994), 23.

140 *"To all whom it may concerne":* In Paul Boyer and Stephen Nissenbaum, eds. *The Salem Witchcraft Papers: Verbatim Transcripts of the Legal Documents of the Salem Witchcraft Outbreak of 1692*, vol, 2 (New York: De Capo Press, 1977), 592.

140–41 *"Moast Grand wise":* Ibid., 607.

141 *"the honoured court":* Ibid.

141 *"that poisoned cloud":* Arthur Miller, "Why I Wrote the 'Crucible,'" *New Yorker*, October 21 and 28, 1996, 164.

143 *"to Complain of":* In Trask, op. cit., 50.

145 *"Being chain'd in the dungeon":* Ibid., 125.

145 *"all might be covered":* In Starkey, op. cit., 251.

145 *"each generation must":* Richard B. Trask, ed., *Danvers Remembers: The Commemoration of the Tercentennial of the 1692 Salem Village Witchcraft Delusion* (Danvers, Mass.: Salem Village Tercentennial Committee, 1993), 3.

Julia Morgan's Hearst Castle, San Simeon, California

146 *"Dear Miss Morgan":* Quoted in Nancy E. Loe, *Hearst Castle: An Interpretive History of W. R. Hearst's San Simeon Estate* (Santa Barbara, Calif.: Companion Press, 1998), 17.

146 *"did not want to encourage":* Ibid., 18.

146 *"the best and most talented":* Ibid.

146 *"He could make or break you":* Ibid., 29.

146 *"something that would be":* Ibid., 8.

146 *"The experts told Pop":* Ibid.

148 *"brought from the ends":* Ibid., 48.

148 *"We are building":* In Sara Holmes Boutelle, *Julia Morgan, Architect* (New York: Abbeville Press, 1988), 184.

148 *"The big house is a whale":* In Loe, op. cit., 28.

149 *"the longest in captivity":* In Robert L. Pavlik: "Something a Little Different," *California History Magazine*, Winter 1992–1993, 475.

149 *"furnished with antiques":* Boutelle, op. cit., 198.

149 *"This is the way God":* Phyllis Theroux, "No Place Like Home: Two American Palaces," *New York Times Magazine*, October 20, 1991, 25.

149 *hiring, firing:* Boutelle, op. cit., 212–213.

150 *"a neat bantam hen":* Boutelle book review, *New York Times*, March 28, 1988.

150 *"money was a small part":* In Boutelle, op. cit., 47.

152 *"Mr. Hearst and I":* In Roger W. Moss, *The American Country House* (New York: Henry Holt & Co.: 1990), 208.

152 *"Everyone was looking":* The Julia Morgan Architectural History Project, "Three conversations with Morgan and Flora North about Julia Morgan" (the Regents of the University of California, 1976), 204.

152 *"They were both long distance":* In Taylor Coffman, *Hearst Castle: The Story of William Randolph Hearst and San Simeon* (Sequoia Books, ARA Leisure Services, and Taylor Coffman), 46.

152 *I like your idea:* In Loe, op. cit., 69.

152 *"probably a windfall":* North conversations, 180.

152 *"But these are":* In Boutelle, op. cit., 116.

155 *"a burning disgrace":* In W. A. Swanberg, *Citizen Hearst: A Biography of William Randolph Hearst* (New York: Charles Scribner's Sons, 1961), 246, n. 13.

155 *"the publisher":* Ibid., 246.

155 *"Heartily approve":* In Boutelle, op. cit., 188.

155 *"He just lived for plans":* Marion Davies, *The Times We Had: Life with William Randolph Hearst* (Indianapolis and New York: Bobbs-Merrill Co. 1975), 45.

155 *"to be able to build":* North conversations, 220.

155 *"this great Californian":* In Loe, op. cit., 8.

Eleanor Roosevelt's Val-Kill Cottage, Hyde Park, New York

156 *"somewhat odd":* Eleanor Roosevelt, *The Autobiography of Eleanor Roosevelt* (New York: Da Capo Press, 1992), 145.

156 *"perfectly good":* "Out from Under Sara's Thumb," *Barrytown Explorer*, November 1980 (FDR Library).

156 *"fairly comfortable":* Roosevelt, op. cit., 145.

156 *"The peace of it":* In Geoffrey Ward, "Eleanor Roosevelt at Val-Kill," *Smithsonian* 15, no.7 (October 1984), 66.

156 *"I could be happy":* In *The Val-Kill Years*, National Park Service brochure.

156 *"She loved having company":* Marguerite Entrup oral history, January 10, 1979, for the Franklin D. Roosevelt Library, 1.

156 *"only 14 tomorrow":* In Ward, op. cit., 71.

156 *"My mother-in-law":* Roosevelt, op. cit., 293.

156 *"She could not bear":* In Geoffrey Ward, op. cit., 70.

156 *"Possessions seemed":* In *Historic Furnishings Report, Home of FDR (Springwood)*, vol. 1 National Historic Site, Hyde Park, New York, 16.

158 *"Driving thro'":* In Joseph P. Lash, *Love, Eleanor: Eleanor Roosevelt and Her Friends* (Garden City, N.Y.: Doubleday & Co., 1982), 208.

158 *"She took you":* Entrup oral history, op. cit., 24.

158 *"There isn't anybody":* Author interview with Nina Gibson, May 1997.

158 *"I'm always given":* In Joseph P. Lash, *Eleanor and Franklin* (New York: W. W. Norton & Co., 1971), 475.

158 *"all of us need":* Ibid.

159 *"Perhaps we need":* In Lash, *Love, Eleanor*, 151.

159 *"I like the still nights":* In Joseph P. Lash, *A World of Love: Eleanor Roosevelt and Her Friends, 1943–62* (Garden City, N. Y.: Doubleday & Co., 1984), 243.

159 *"Dear God":* In Doris Kearns Goodwin, "The Home Front," *New Yorker*, August 15, 1994, 41.

160 *"Human beings":* In Lash, *Love, Eleanor,* 181.

160 *"You never sent":* Entrup oral history, op. cit., 15–16.

160 *"we do not have to become":* In Geoffrey C. Ward,"Mrs. Roosevelt Faces Fear," *American Heritage,* October 20, 1984, 19.

162 *"the two homeliest people":* In Lash, *Love, Eleanor,* xiv.

162 *"extraordinary eyes":* Roosevelt, op. cit., 109.

162 *"He always wanted":* In Lash, *Love, Eleanor,* 278.

162 *"What a nuisance":* Ibid., 219.

162 *"My husband and I":* Roosevelt, op. cit., 283.

162 *"not to accept":* In Doris Kearns Goodwin, *No Ordinary Time: Franklin and Eleanor Roosevelt: The Home Front in World War II* (New York: Simon & Schuster, 1994), 338.

163 *"I remember":* In Geoffrey C. Ward, *Before the Trumpet, Young Franklin Roosevelt, 1882–1915* (New York: Harper & Row, 1985), 319.

163 *"Through the whole":* Roosevelt, op. cit., 159.

163 *"We have to prove":* In Joseph P. Lash, *The Years Alone* (New York: W. W. Norton & Co., 1972), 154.

163 *"Where after all":* Ibid., 81.

VI: Castles in the Sand

Biltmore Estate, Asheville, North Carolina

166 *"Now I have brought you":* Frederick L. Olmsted letter to Fred Kingsbury, January 20,1891 (facsimile). Olmsted Collection, Library of Congress.

166 *"The woods":* Ibid.

166 *"delicate, refined":* Ibid.

166 *"Such land in Europe":* Ibid.

168 *"Make a small park":* Ibid.

168–70 *"The mountains":* In Paul R. Baker, *Richard Morris Hunt* (Cambridge, Mass.: MIT Press, 1980), 414.

170 *[his] plans for the forest:* John M. Bryan, *Biltmore Estate: The Most Distinguished Private Place* (New York: Rizzoli International Publications, and the American Architectural Foundation, 1994), 35.

171 *"spending more money":* In Arthur Vanderbilt II, *Fortune's Children: The Fall of the House of Vanderbilt* (London: Michael Joseph, 1989), 299.

171 *"to give it":* In Bryan, op. cit., 32.

171 *so "the water":* In Bryan, op. cit., 94.

171 *"with an abrupt":* Ibid., 34.

171 *"would give Olmsted":* In *A Guide to Biltmore: Special Centennial edition* (Asheville, N. C.: Biltmore Co., 1994), 81.

172 *"one long tale":* In Baker, op. cit., 431.

172 *"castle of enchantment":* In Louis Auchincloss, *The Vanderbilt Era: Profiles of a Gilded Age* (New York: Charles Scribner's Sons, 1989), 60.

172 *"glacial phantasy":* In Baker, op. cit., 431.

172 *"bloated Biltmore":* In Auchincloss, op. cit., 61.

172 *so he couldn't have:* George Washington Vanderbilt's trip diary, July 30, 1880, bound volume, Biltmore Estate Archives, Asheville, North Carolina.

172 *"the Great Model Estate":* George F. Weston, "Biltmore," in *Country Life Magazine,* September, 1902, 180.

173 *"The trouble with you . . . architects":* In John Foreman and Robbe Pierce Stimson, *The Vanderbilts and the Gilded Age: Architectural Aspirations, 1879–1901* (New York: St. Martin's Press, 1991), 296.

173 *"If we had known":* In Laura Wood Roper, *FLO: A Biography of Frederick Law Olmsted* (Baltimore and London: Johns Hopkins University Press, 1973), 477.

173 *"shrewd, sharp":* FLO letter to Fred Kingsbury, January 20, 1891.

173 *I have been way:* GWV's diary, May 27, 1875, bound volume, Biltmore Estate Archives, Asheville, North Carolina.

173 *"casual and informal":* In Jerry Patterson, *The Vanderbilts* (New York: Harry N. Abrams, 1989), 188.

174 *"a devastating commentary":* Ibid., 184.

174 *"happy as a clam":* In Clive Aslet, *The American Country House* (New Haven and London: Yale University Press, 1990), 14.

174 *"a work of very rare public interest":* FLO letter to William H. Thompson, November 6, 1889, Olmsted Collection, Library of Congress.

174 *"courteous in manner":* Biltmore exhibition. text.

174 *"It is a slice of history":* Author interview with William A. V. Cecil, January 18, 1995.

Lower East Side Tenement Museum, New York, New York

176 *The crush and the stench:* In Irving Howe, *World of Our Fathers: The Journey of the East European Jews to America and the Life They Found and Made,* abridged ed. (New York: Bantam Books, 1980), 64.

178 *"where the greatest":* In Andrew Dolkart, "97 Orchard Street: The Tenement Museum and Tenement House Reform, An Architectural Analysis" (unpublished paper, 1995), 5.

178 *"The structure":* In Leslie Mieko Yap, "Talk of the Tenement," *Modern Maturity* 34, no.2 (1991), 4–5.

180 *"As soon as he can":* Jacob Riis, "How the Other Half Lives," *in Jacob Riis Revisited: Poverty and the Slum in Another Era* (Garden City, N. Y.: Anchor Books/Doubleday, 1968), 24.

180 *"watershed moment":* Gary Kulik, *New York Teacher*, City ed., March 7, 1994, 13A.

183 *"as if they were made":* Riis, op. cit., 326.

183 *"plainly due to suffocation":* Ibid., 10.

184 *"the people didn't":* Sam Bass Warner, editor's introduction to *How the Other Half Lives: Studies Among the Tenements of New York* (Cambridge, Mass.: Belknap Press of Harvard University, 1970), xiii.

184 *How well I remember:* all quotes here from the Lower East Side Tenement Museum Memory Board.

'Iolani Palace, Honolulu, Hawai'i

186 *"filthy and in poor condition":* Richard A. Wisniewski, *Hawaiian Monarchs and Their Palaces: A Pictorial History* (Honolulu: Pacific Basin Enterprises, 1987), 52.

186 *"put a girdle":* In William N. Armstrong, *Around the World with a King* (Rutland, Vt., and Tokyo, Japan: Charles E. Tuttle Co., 1997), 2.

186 *"so that we may become":* In Terence Barrow, ibid., Introduction, xxiii.

186 *... while he was working:* Lili'uokalani, *Hawai'i's Story by Hawai'i's Queen* (Honolulu: Mutual Publishing, 1990), 77.

188 *"even bungled":* In Gavan Daws, *Shoal of Time: History of the Hawaiian Islands* (New York: Macmillan Co., 1968), 219.

188 *"At the moment":* Lili'uokalani, op. cit., 103–105.

189 *"Americans own":* Mark Twain: *Letters from Hawai'i*, edited and with an introduction by A. Grove Day (London: Chatto & Windus, 1967), 132.

189 *"Kalākaua's reign":* Lili'uokalani, op. cit., 233.

193 *"by the base ingratitude":* Ibid., 192.

194 *"just emerging":* In S.J. Resolution 19, 5. Accompanying document. 103rd Cong. 1st Sess. Calendar #185, Report #103.126, 2.

194 *"sophisticated language":* Ibid.

194 *That first night:* Lili'uokalani, op. cit., 269.

195 *"with the gaze of strangers":* Ibid., 317.

195 *"the true Isles of the Blest":* In A. Grove Day, introduction to *Letters from Hawai'i*, by Mark Twain, xiv.

VII: Haunted Houses

Edgar Allan Poe's Cottage, Bronx, New York

197 *"genius of American terror":* Mark Edmundson, *Nightmare on Main Street: Angels, Sadomasochism, and the Culture of Gothic* (Cambridge, Mass., and London, England: Harvard University Press, 1997), 71.

198 *"I intend":* In Kenneth Silverman, *Edgar A. Poe: Mournful and Never-Ending Remembrance* (New York: HarperCollins Publishers, 1991), 145.

198 *"the Comanche of literature":* In Dwight Thomas and David Jackson, eds., *The Poe Log:A Documentary Life of Edgar Allan Poe, 1809–1849* (Boston: G. K. Hall & Co., 1987), 663.

198 *There could be nothing:* From Edgar Allan Poe's "Philosophy of Furniture," in *House & Garden*, November 1983, 158ff. and passim.

198 *"a footman":* Silverman, op. cit., 7.

198 *"It is an evil":* Poe, op. cit., 210.

200 *"old and buggy":* In Philip Van Doren Stern, ed. *The Portable Poe* (New York: Penguin Books, 1945), 19.

200 *"a small wooden box":* In Mary E. Phillips, *Edgar Allan Poe: The Man*, vol. 2 (John C. Winston Co., 1926), 1113.

201 *"the ludicrous":* In Silverman, op. cit., 110.

201 *"Let him remember":* In Thomas and Jackson, op. cit., 678.

201 *"darling little wifey,":* In Stern, op. cit., 15.

201 *Ever with thee:* In Elizabeth Beirne, ed., introduction to *Poems of Edgar Allan Poe at Fordham* (New York: Bronx County Historical Society, 1980).

201 *"half buried in fruit":* In Thomas and Jackson, op. cit., 639.

202 *"It was the sweetest":* In Henry Noble McCracken, "Poe's Life at Fordham," *Transactions of the Bronx Society of Arts and Science* 1, pt. 2 (May 1910), 26.

202 *"to battle with":* In Phillips, op. cit., 1118.

202 *"She loved to sit":* Ibid., 1183.

202 *[It] had an air:* Mary Gove Nichols, *Reminiscenses of Edgar Allan Poe* (New York: Union Square Book Shop, 1931), 89.

202 *"They were awful poor":* In Phillips, op. cit., 1121.

202 *"We knew the sadness":* Ibid., 1182.

203 *"Mrs. Poe sank rapidly":* In Nichols, op. cit., 12.

203 *"with long intervals":* In Silverman, op. cit., 334.

203 *"I see no one":* In Phillips, op. cit., 1184.

205 *"I should, both as a physician":* In Peggy Robbins, "Poe's Defamation," in *Edgar Allan Poe, The Creation of a Reputation* (Harrisburg, Pa.: National Historical Society and Eastern Acorn Press, 1983), n.p.

205 *"He was his own":* In Nichols, op. cit., 14.

205 *"symbolic confessional":* In Stern, op. cit., 56, 82.

205 *"and he was not responsible":* In Silverman, op. cit., 184.

205 *"It has not been":* In Stern, op. cit., xx.

205 *"his clear and vivid perception":* In Nichols, op. cit., 14.

205 *"his devotion":* Jeffrey Meyers, *Edgar Allan Poe: His Life and Legacy* (New York: Charles Scribner's Sons, 1992), 262.

205 *"In bearing":* Ibid., 264.

206 *"force and originality":* Ibid.

206 *"that jingle man":* Ibid.

206 *"literary glory of America":* Poe Museum, Richmond, Va.

206 *"My personal debt":* In Meyers, op. cit., n. 291.

206 *"Where was the detective story":* *Four Faces of Poe*, National Park Service Teacher's handbook, n.d., n.p.

206 *"So many things": Philadelphia Inquirer*, April 18, 1983.

206 *"Poe was ... a poor devil":* In Meyers, op. cit., 271.

Janet Sherlock Smith's South Pass Hotel, South Pass City, Wyoming

208 *South Pass Hotel!:* In *South Pass News*, a Publication of the South Pass City State Historical Site.

208 *We overtook:* Mark Twain, *Roughing It* (New York:Harper & Bros., 1871–1899), 81.

212 "Think of hotel-keeper": Ibid., p. 83.

212 *"We have had from 2 to 15":* Todd Guenther, ed. "Dear Peter: The Letters of a Pioneer Mother and Sister," *Wind River Mountaineer* 7, no. 2 (April–June 1991) (published by the Fremont County Museum, Wyoming), 14.

213 *"We have been busy":* Ibid., 18.

213 *"They make as much fuss":* Ibid., 19.

215 *"We got her home":* Ibid., 23.

216 *"We were so happy":* Ibid., 25.

216 *"There is one thing":* Ibid., 19.

216 *"I'm getting tired":* Ibid., 27.

216 *I have lived here:* In Todd Guenther, *A Short Biography of Janet McComie Sherlock Smith, 1844–1923*, review draft, p. 37.

Edward Gorey's House, Yarmouth Port, Massachusetts

218 *fun to do Poe:* Tim Wood, "Frightening Things About Life, According to Edward Gorey," *A-Plus Magazine*, October 1995 (published in Yarmouth Port, Mass.).

218 *"Life is basically ghastly":* Quoted in Alan W. Petrocelli, "A Niche of His Own," *Prime Time Cape Cod* (Hyannis, Massachusetts), October 1995.

218 *"useful for satire":* Wood, op. cit.

218 *"In Gorey's world":* D. Keith Mano, Edward Gorey profile, *People Magazine*, July 3, 1978, 73.

218 *"It's impossible to know":* In Stephen Schiff, "Edward Gorey and the Tao of Nonsense," *New Yorker*, November 9, 1992, 89.

221 *"There is an order here":* Author interview with Edward Gorey, April 6, 1985.

222 *"so I don't need much room":* In Cliff Henderson, "E. Is for Edward Who Draws in His Room," *Arts and Entertainment Magazine,* October 1991.

222 *"I would rather not":* In Clifford Ross and Karen Wilkin, *The World of Edward Gorey* (New York: Harry N. Abrams, 1996), 20.

222 *"a great way":* Wood, op. cit.

223 The Glorious Nosebleed: *Edward Gorey, The Glorious Nosebleed: Fifth Alphabet* (New York: Dodd Mead & Co., 1975).

VIII: Plain and Fantasy

Cedric Gibbons/Dolores Del Rio house, Santa Monica, California

226 *"He knew what to put":* Billy Wilder portrait, *Architectural Digest,* April 1994, 26.

226 *"it was the only cozy place":* Quoted in Charles Lockwood, *Dream Palaces: Hollywood at Home* (New York: Viking Press, 1981), 184.

229 *"Nothing, absolutely nothing":* In Michael Webb, "Cedric Gibbons and the MGM Style," *Architectural Digest,* April 1990, 100.

229 *"Hollywood's most amazing home":* In Robert Heide and John Gilman, *Popular Art Deco: Depression Era Style and Design* (New York: Abbeville Press, 1991), 200.

229 *"a stucco that went to college":* Author telephone interview with Avery Faulkner, architect, Washington, D.C., n. d.

230 *"like an Aztec princess":* Errol Flynn, *My Wicked, Wicked Ways* (New York: G. P. Putnam's Sons, 1959), 192.

230 *"aching to play":* In Heide and Gilman, op. cit., 202.

230 *"Whatever you put there":* In Webb, op. cit., 104.

230 *"the more frequently":* Cedric Gibbons in *Encyclopaedia Britannica,* 14th ed. vol. 24, 858.

232 *"just a plain comfortable":* In Lockwood, op. cit., 54.

233 *"If realism can be abandoned":* Gibbons, op. cit., 860.

233 *"I believe that Romance":* In Richard Striner, "Art Deco: Polemics and Syntheses," *Winterthur Portfolio,* Spring 1990, 27.

Vanna Venturi House, Philadelphia, Pennsylvania

234 *[using] a joke:* Robert Venturi, Denise Scott Brown, Steve Izenour, *Learning from Las Vegas,* rev. ed. (Cambridge, Mass., and London: MIT Press, 1977), 161.

234 *"In its sheltering manner":* Quoted in Frederick Schwartz, ed., introduction to *Mother's House: The Evolution of Vanna Venturi's House in Chestnut Hill* (New York: Rizzoli International Publications, 1992), 37.

234 *"it's perhaps the most important":* Ibid.

234 *"It hurts":* Schwartz, Ibid., 35.

234 *"walked on air":* Ibid., 19–20.

234 *"not by habit":* Robert Venturi, *Complexity and Contradiction in Architecture,* 2nd ed. (Museum of Modern Art in association with the Graham Foundation for Advanced Studies in the Fine Arts, Chicago, distributed by Harry N. Abrams, 1977), 13.

234 *"the variety with the vulgarity":* Venturi et al., *Learning,* 153.

234 *"Less is a bore":* In Alexander Boulton; "On Mother's House," *American Heritage,* July–August 1996, 99.

234 *"sound but unorthodox":* In Schwartz, op. cit., 17.

234 *"This facade":* In Boulton, op. cit., 95.

236 *"holes in the wall":* Venturi, *Complexity,* 118ff.

236 *"You learn a lot":* Venturi lecture on videotape, March 1997.

237 *"I am for messy vitality":* Venturi, *Complexity,* 16.

237 *"it's not a solemn house":* Author interview with Agatha Hughes.

238 *"puppy with large feet":* Denise Scott Brown, *Architectural Monographs,* no. 21 (London: Venturi, Scott Brown & Associates, and the Academy Group, 1992), 13

239 *"Experts with Ideals":* Venturi, Scott Brown, and Izenour, op. cit., 154.

239 *"Main Street is almost":* Quoted in Vittorio Lampugni, "Looking at Architecture with New Eyes," *Pritzger Architecture Prize presented to Robert Venturi* (Los Angeles: Jensen & Walker, Inc., 1991).

239 *"Landscapes should include":* Robert Venturi, "Diversity, relevance and representation in historicism, or plus ça change," *Architectural Record*, June 1982, 115.

239 *"The Italian landscape":* Venturi, et al., *Learning*, 6.

239 *"Sprawl and strip":* Ibid., 155.

239 *"they have changed":* Paul Goldberger, *New York Times* August 14, 1991.

Hariri & Hariri's House for the Next Millennium, Exit 2000, United Expressway 21

240 *I dwell in possibility:* Thomas H. Johnson, ed., *The Complete Poems of Emily Dickinson* (Boston: Little, Brown & Co., 1960), 327.

240 *"They exemplify": Hariri & Hariri*, essays by Kenneth Frampton and Steven Holl, compiled by Oscar Riera Ojeda (New York: Monacelli Press, 1995), front flap of book jacket.

240 *"a testing ground":* Ibid., 140.

240 *"By dream":* In D. S. Friedman, "Wild Walls," in *The Architect's Dream: Houses for the Next Millennium* (Cincinnati: Contemporary Arts Center, 1993), 8.

243 *"We need a totally new outlook":* In Friedman, op. cit., 24.

243 *"A family of four":* Ibid., 24.

244 *"metaphorically":* Author telephone interview with Gisue Hariri, May 21, 1997.

244 *"find out what people": Harper's Bazaar*, September 1995, 352.

244 *"They are spaces": Hariri & Hariri*, 70.

244 *"Here you will face": Harper's Bazaar*, op. cit.

246 *"The contrast of materials":* Ibid., 37ff.

246 *"like thick, solid stone": Hariri & Hariri*, 33–34.

246 *"We carry a lot of memories":* Ibid., 26.

246 *"single sheet":* Ibid., 100.

246 *"housing for the schizophrenic":* Ibid., 91.

246 *"Now that we are": Harper's Bazaar*, op. cit.

246 *"a busy woman":* H. Ward Jandl, *Yesterday's Houses of Tomorrow: Innovative American Homes 1850 to 1950* (Washington, D. C.: Preservation Press, National Trust for Historic Preservation, 1991), 149.

247 *"the human desire":* Jayne Merbel, "Reaching Out: Hariri & Hariri," *Oculus* 57, no. 9 (May 1995), 10.

ACKNOWLEDGMENTS

To Jacqueline Kennedy Onassis, whose encouraging words as this book got underway—"I think it's thrilling and inspiring"—brightened many a day.

To the four people whose contributions were vital to the making of the book:

Jane S. Caplan, skilled interviewer and researcher, who could always find anyone and everything needed; intrepid traveling companion who always got us to the house on time, and whose dedication to this project is unsurpassed.

Paul Mahon, literary agent, warm and generous guide through unfamiliar publishing territory.

Ann Monroe Jacobs, meticulous and determined image researcher whose work enriches the text so beautifully.

Paul Johnson, courtly supporter of a "compelling" idea.

To those who provided financial support at critical stages in our research and development: the American Association of Housing Educators and Dr. Joseph Wysocki, Friends of the Vann House, the Graham Foundation for Advanced Studies in the Fine Arts, the LBJ Family Foundation, George Mitchell, the L. J. Skaggs and Mary C. Skaggs Foundation, the James Smithson Society, and the Smithsonian Women's Committee.

To those experts consulted at the beginning, to provide the broadest sweep of prospective houses: Clive Aslet, Dr. James Barber, Peggy Berryhill, Pete Daniel, Peter Decker, Dr. Wilton Dillon, Dr. Alan Fern, Dr. Cynthia Field, Mary Mix Foley, Timothy Foote, Tom Freudenheim, Elizabeth Garrett, the late Brendan Gill, Dr. Alicia Gonzalez, Dr. Rayna Green, Janet Heller, Dr. Thomas Hoving, Ellen Roney Hughes, Enid Hyde, James Loewen, Dr. Chauncey Loomis, the late Russell Lynes, Dr. Mina Marefat, Constance Martin, David McCullough, David McFadden, Dr. Keith Melder, the Miltons, Dr. Arthur Molella, Dr. Barbara Mossberg, Marc Pachter, Ford Peatross, Buck Pennington, Dr. Dwight Pitcaithley, Aleta Ringlero, Andrea Schmertz, Mildred Schmertz, Dr. Theresa Singleton, Paul Stewart, Michael Webb, and Dr. Gwendolyn Wright.

To the scholars, curators, and guides at specific sites, who were so consistently generous with their welcomes, advice, and corrections:

CHAPTER ONE: LIVING IN ART—Pat Baillargeon, Marge Byrd and members of the Wrangell IRA Tribal Council, Dawn Hutchinson, Mary Kowalczyk, Nathan Jackson, Luree Miller, Lilian Prevet, Carol Rushmore, Margaret Sturtevant, and Joe Williams; Gordon Alt, Jonathan Leck, Joel Silver, and Eric Lloyd Wright; Dr. Edward Bosley, Sally Gamble Epstein, and James Gamble; Dr. Lesley Lee Francis, Robin Hudnut, Dr. Edward

Connery Lathem, Dr. Donald Sheehan, and Claire Ternan; Dr. Charles Eldredge; Mikka Gee, Elizabeth Glassman, and Judy Lopez.

CHAPTER TWO: GEORGE WASHINGTON DIDN'T SLEEP HERE—Dr. Joseph Ellis and Dr. Alan Kraut; Caroline Keinath, Marianne Peak, and Celeste Walker; Larry Dermody, Randy Huwa, Lee Langston-Harrison, Lynne Lewis, Marcy Modeland, Dr. William Lee Miller, and Dr. James Morton Smith; Kim Bauer, Dr. Cullom Davis, Mary Ellen McElligot, and Linda Norbut-Suits; Dr. Bernard Bailyn, Skip Cole, Ann Jordan, Dr. Thomas Mann, and Dr. Pauline Maier.

CHAPTER THREE: THE TRUTH ABOUT TARA—Dr. James O. Horton and Dr. Charles Joyner; Dr. John Hope Franklin and John Franklin, Portia James, Dr. Faith Ruffins; Will Mangham, Gene Slivka, Miss Mamie Thompson, and Dr. Suzanne Turner; Donna Fricker, Dr. Daniel Schafer, Kathy Tilford, and Dr. John Michael Vlach; Ezby Collins, Ken Heidelberg, Saundra Jackson, Janice McGuire, and Barbara Tagger; Betty Campbell, Lobena Frost, Charles Nuckolls, Stuart Seely Sprague, and Miriam Zachman; Reverend K. G. Jones; Kay Montgomery, Payne and Harrison Tyler; Arlin Dease, Dr. Sydney Nathans, Dorothy Redford, Mary Vance Trent, and Dr. Peter Wood.

CHAPTER FOUR: FORGOTTEN FRONTIER—Sue Fisher, Ramson Lomatewama, Todd Metzger, and David Noble; Julia Autry, Pat and Ed Hall, Tim Howard, and Jeff Stancil; Larry Costa, Earl Douglass, and James B. Alexander; David Brugge and Nancy Stone; Jane Russell and Dr. Herbert Yee; Steade Craigo, Eugene Itogawa, Connie King, Ping Lee, Elizabeth Olds, Carl and Carol Sandmeier, and Lonn Taylor.

CHAPTER FIVE: A WOMAN'S PLACE—Richard Trask; James Allen, Sara Boutelle, and Evelyn Gardiner; Maureen Corr, Nina Gibson, Franceska Macsali, Jane Plakias, and Eleanor Seagraves; Ann Brittain, Lesley and Todd Cooper, Terrel Delphin, Dr. Hiram Gregory, Betty Metoyer, Maxine Southerland, Mary Linn Wernet, and Thomas Whitehead.

CHAPTER SIX: CASTLES IN THE SAND—Cathleen Baldwin, Rick King, Jerry Patterson, Pat Rojas, and Elizabeth Sims; James Bartels, Dr. Gavan Daws, Corinne Chun Fujimoto, and Rich Kennedy; Ruth Abram, Anita Jacobson, Steve Long, and Edward Dolkart; Marcia Stout.

CHAPTER SEVEN: HAUNTED HOUSES—Kathleen McAuley, Jeff Jerome, and Regina Underwood; Todd Guenther, Scott Goetz, and Tom Lindmier; Grace Budd and Edward Gorey; Fenwick Kollek, Brad Sturdevant.

CHAPTER EIGHT: PLAIN AND FANTASY—Dianne Pilgrim; Martin Weil, Julius Shulman, Adele and Ira Yellin; Thomas and Agatha Hughes; Dr. Barbara Kelly, Louise Cassano, Lynn Matarrese, Roberta Ross, Ralph and Sally Schneider; Joshua Soren and Dr. Harold Wattel; Avery Faulkner, Mojgan and Gisue Hariri; James Cutler, Clifford Pearson, and Stephen Perella.

To Scott Moyers, who first brought me to Simon & Schuster, and to Sharon Gibbons, my editor, and William Rosen, editorial director, who have so ably and agreeably made this book a reality.

To Christopher and Susan Koch, distinguished producers who will turn this book into a television series one day.

To the Smithsonian Library staff for their first-rate assistance: Claire Catron, Jim Harar, Polly Lasker, Amy Levin, Rhoda Ratner, Martin Smith, Stephanie Thomas, and Wanda West, and lastly:

To my family and friends, whose faithful question "How's the book coming?" spurred me on, especially to Alice Smith, Adrian Malone, and Robert Immerman, whose remark, "Have I got a house for you!" led to some interesting discoveries.

Any list with every name of those we have called on for advice or assistance during the course of this project would be too long, but the memory of *all* those we have met along the way remains vivid, and to them we are also deeply grateful.

Melrose Plantation, near Natchitoches, Louisiana, is one of the many remarkable sites visited during the research of this book. This painting, by noted folk artist Clementine Hunter, one-time field hand and plantation cook, shows the three houses on the plantation that together tell her story and that of the two other remarkable women who lived there over three centuries.

PHOTO CREDITS

ii: Courtesy of the Franklin D. Roosevelt Library

vi, top: National Anthropological Archives, Smithsonian Institution (detail)

vi, second from top: U. S. Dept. of the Interior, National Park Service, Adams National Historic Site (detail)

vi, second from bottom: Courtesy of Gene Slivka (detail)

vi, bottom: © John Elk III (detail)

vii, top: Richard B. Trask (detail)

vii, second from top: Fred Hirschmann (detail)

vii, second from bottom: The Bronx County Historical Society Collections (detail)

vii, bottom: Tim Street-Porter (detail)

I: Living in Art

x–1: National Anthropological Archives, Smithsonian Institution

Tlingit Clan House

3: Fred Hirschmann

4: Fred Hirschmann

5, above: Fred Hirschmann

5, below: Barry Herem

6, top: Arctic Studies Center, National Museum of Natural History, Smithsonian Institution (#313279)

6, above: Arctic Studies Center, National Museum of Natural History, Smithsonian Institution (#60144, #60147)

7: National Anthropological Archives, Smithsonian Institution

8: © Ivan Simonek, Alaskan Images

Storer House

10: © 1999 The Frank Lloyd Wright Foundation

11: Balthazar Korab

12, above: Courtesy of The Frank Lloyd Wright Archives

12, right: Tim Street-Porter

13: Tim Street-Porter

14–15: Tim Street-Porter

Auldbrass Plantation

16: Paul Rocheleau

17: Paul Rocheleau

18, above: Paul Rocheleau

18, right: Paul Rocheleau

19: Paul Rocheleau

The Gamble House

20: Tim Street-Porter

21: Tim Street-Porter

22: Photograph © Peter Aaron / ESTO

23: Photograph © Peter Aaron / ESTO

24: Tim Street-Porter

25, top: Tim Street-Porter

25, above: Theodore Thomas

The Robert Frost Farm

27: Eliot Cohen

28, above: Dartmouth College Library

28, below: © John Wells / New England Stock Photo

29: Chris Ternan

30: © John Wells / New England Stock Photo

31, top: Dartmouth College Library

31, below: Courtesy of Lesley Lee Francis

32, above: © John Wells / New England Stock Photo

32, below: © John Wells / New England Stock Photo

34–35: Paul Rocheleau

II: George Washington Didn't Sleep Here

36–37: U.S. Dept. of the Interior, National Park Service, Adams National Historic Site

The Adams Family's House

38–39: © 1999 Joshua Greene, all rights reserved (541) 997-4970

40: U.S. Dept. of the Interior, National Park Service, Adams National Historic Site

41: © 1999 Joshua Greene, all rights reserved (541) 997-4970 (detail)

42: © 1999 Joshua Greene, all rights reserved (541) 997-4970

43: U.S. Dept. of the Interior, National Park Service, Adams National Historic Site

45: © 1999 Joshua Greene, all rights reserved (541) 997-4970

46, left: U.S. Dept. of the Interior, National Park Service, Adams National Historic Site

46, right: U.S. Dept. of the Interior, National Park Service, Adams National Historic Site

47: U.S. Dept. of the Interior, National Park Service, Adams National Historic Site

48–49: © 1999 Joshua Greene, all rights reserved (541) 997-4970

Montpelier

52: Library of Congress

52–53: © Robert Llewellyn

54: Robert Lautman

56–57: Photography by Richard Cheek for the Garden Club of Virginia

57: © Ron Blunt Photography

58: © Ron Blunt Photography

59: Colonial Williamsburg Foundation

61: Courtesy of Montpelier, National Trust for Historic Preservation

Abraham Lincoln's House

62: Courtesy of the Illinois State Historical Library

63: Balthazar Korab

64: Courtesy of the Illinois State Historical Library

65: Library of Congress

66: Balthazar Korab

67: Lincoln Home National Historic Site

68: Lincoln Home National Historic Site

70: Lincoln Home National Historic Site

III: The Truth About Tara

72–73: Courtesy of Gene Slivka

Rosedown Plantation

75: © Tom Till

76: © Collection of The New-York Historical Society

77: Paul Rocheleau

78: Paul Rocheleau

79: Paul Rocheleau

81: Paul Rocheleau

82: Courtesy of Elizabeth Brownstein

84: Courtesy of Gene Slivka

84–85: Paul Rocheleau

Underground Railroad Houses

87: Eastman Johnson's *A Ride for Liberty—The Fugitive Slaves* (#40.59.A), courtesy of The Brooklyn Museum of Art

88, near right: © Rod Berry

88, far right: The John P. Parker Historical Society, Courtesy of Office of Imaging, Printing, and Photographic Services, Smithsonian Institution

89: © Louie Psihoyos / Contact Press Images

90, top: Ohio Historical Society

90, above: Ohio Historical Society

91: © Rod Berry

92: © Rod Berry

93: Ohio Historical Society

The Oaks

95: © R. Hagerty, courtesy of Tuskegee Institute National Historic Site

96: Joe Zentner

97: Library of Congress

98: Photograph by Jonathan Wallen ©

99: © R. Hagerty, courtesy of Tuskegee Institute National Historic Site

100: © Eastern National, courtesy of Tuskegee Institute National Historic Site

101: Tuskegee University Archives

IV: Forgotten Frontier

102–103: © John Elk III (detail)

Wukoki Pueblo

104: © Jerry Jacka, 1999

104–105: © Jerry Jacka, 1999

106: David Grant Noble

107, above: Library of Congress

107, below: David Grant Noble

109: Photograph by Eric Swanson, courtesy Sakiestewa Textiles Ltd. Co., The Ancient Blanket Series—"Wupatki"

Cherokee Chief Vann's House

111: Elvin Strickland

113, top: Elvin Strickland

113, bottom: Elvin Strickland

115: Elvin Strickland

116, above: Woolaroc Museum, Bartlesville, Oklahoma

116, below: Courtesy of National Museum of the American Indian, Smithsonian Institution (05/0770)

117, above left: Courtesy of Chief Vann House State Historic Site, Friends of the Vann House

117, above: Elvin Strickland

Petaluma Adobe

118–19: © John Elk III

121, above: © John Elk III

121, left: © John Elk III

123: © John Elk III

125: The Society of California Pioneers

126: © California State Parks, 1999

Lorenzo Hubbell's House

128: © John Elk III (detail)

128–129: George Huey

130: Courtesy of the National Park Service

131: Fred Hirschmann

132: © Jerry Jacka, 1999

133: George Huey

134: George Huey

135, top: George Huey

135, bottom: Courtesy of the National Park Service, Hubbell Trading Post National Historic Site

V: A Woman's Place

136–37: Richard B. Trask

Rebecca Nurse Homestead

139: Barry S. Kaplan / The Finer Image

140: The Massachusetts Historical Society, Boston

141: Barry S. Kaplan / The Finer Image

142: Barry S. Kaplan / The Finer Image

143, left: Barry S. Kaplan / The Finer Image

143, below: The Granger Collection, New York

144: Barry S. Kaplan / The Finer Image

Julia Morgan's Hearst Castle

146: Special Collections, California Polytechnic State University

147: Hearst San Simeon State Historical Monument TM

148: Hearst San Simeon State Historical Monument TM

149: Marc Wanamaker / Bison Archives

150, right: Hearst San Simeon State Historical Monument TM

150, below: Marc Wanamaker / Bison Archives

151: Hearst San Simeon State Historical Monument TM

153: © C. Seghers / H. Armstrong Roberts, Inc.

154–55: Hearst San Simeon State Historical Monument TM

155: Hearst San Simeon State Historical Monument TM

Val-Kill Cottage

156–57: Photograph by Richard Cheek for the Hyde Park Historical Association

158: Courtesy of the Franklin D. Roosevelt Library

159: Courtesy of the Franklin D. Roosevelt Library

160, top: Courtesy of the Franklin D. Roosevelt Library

160, bottom: Courtesy of the Franklin D. Roosevelt Library

161, top: Photograph by Richard Cheek for the Hyde Park Historical Association

161, bottom: Photograph by Richard Cheek for the Hyde Park Historical Association

162: Courtesy of the Franklin D. Roosevelt Library

VI: Castles in the Sand

164–65: Fred Hirschmann (detail)

Biltmore Estate

166: Courtesy of Biltmore Estate, Asheville, North Carolina

167: Courtesy of Biltmore Estate, Asheville, North Carolina

168, above: Prints and Drawings Collection, The Octagon, Washington, D.C.

168, below: Courtesy of Biltmore Estate, Asheville, North Carolina

169: Photograph by Richard Cheek for Yale University Press

170: Balthazar Korab

171: Courtesy of Biltmore Estate, Asheville, North Carolina

172: Courtesy of Biltmore Estate, Asheville, North Carolina

173: Courtesy of Biltmore Estate, Asheville, North Carolina

175: © 1999 by Tim Buchman/Transparencies, Inc.

Lower East Side Tenement Museum

176–77: Museum of the City of New York, The Byron Collection

178: Library of Congress

179, left: Collection of the Lower East Side Tenement Museum

179, above: Collection of the Lower East Side Tenement Museum

180, above: Collection of the Lower East Side Tenement Museum

180, right: Steve Brosnahan

181: Steve Brosnahan

182: Steve Brosnahan

185: Elizabeth McCausland Papers, Archives of American Art, Smithsonian Institution

'Iolani Palace

186: © The Friends of 'Iolani Palace

186–87: © 1993 The Friends of 'Iolani Palace

191, top: © 1992 Photography by Milroy/McAleer for The Friends of 'Iolani Palace

191, bottom: © 1992 Photography by Milroy/McAleer for The Friends of 'Iolani Palace

192: © 1992 Photography by Milroy/McAleer for The Friends of 'Iolani Palace

193: R. J. Baker Collection, Bishop Museum

194: Balthazar Korab

195, left: © 1994 The Friends of 'Iolani Palace

195. above: © The Friends of 'Iolani Palace

VII: Haunted Houses

196–97: The Bronx County Historical Society Collections

Edgar Allan Poe's Cottage

198–99: Steve Brosnahan

201: Lilly Library, Indiana University, Bloomington, Indiana

203: Steve Brosnahan

204: Steve Brosnahan

206: Library of Congress

206–207: Steve Brosnahan

South Pass Hotel

209: © Greg Ryan/Sally Beyer

210: © 1999 Corinne Humphrey, Photographer

212: South Pass City State Historic Site

213: South Pass City State Historic Site

214: Wyoming Division of Cultural Resources

215: South Pass City State Historic Site

217: South Pass City State Historic Site

Edward Gorey's House

218: © 1999 Edward Gorey

218–19: © Alan Briere

220, above: © Alan Briere

220, right: © Alan Briere

221: © Alan Briere

222: © 1999 Edward Gorey

223: © Alan Briere

VIII: Plain and Fantasy

224–25: Tim Street-Porter

Gibbons/Del Rio House

227: Tim Street-Porter

228–29: Tim Street-Porter

229: Courtesy of The Academy of Motion Pictures Arts and Sciences

231: Tim Street-Porter

232: Tim Street-Porter

233: Tim Street-Porter

Vanna Venturi House

234: Courtesy of Venturi, Scott Brown and Associates, Inc.

234–35: Courtesy of Venturi, Scott Brown and Associates, Inc., photograph by Rollin R. La France

236: © Birgitta Ralston, courtesy of Agatha Hughes

237: © Steven Goldblatt

238: © Steven Goldblatt

239: © Steven Goldblatt

The House for the Next Millennium

241: © Richard Scanlan, courtesy of Mojgan and Gisue Hariri

242: © John M. Hall, courtesy of Mojgan and Gisue Hariri

243, above: Courtesy of Mojgan and Gisue Hariri

243, left: Courtesy of Mojgan and Gisue Hariri

244: © Richard Scanlan, courtesy of Mojgan and Gisue Hariri

245: © John M. Hall, courtesy of Mojgan and Gisue Hariri

246: © Steve Cohen, courtesy of Mojgan and Gisue Hariri

247: The University of New Mexico Art Museum, photograph by Damian Andrus

265: *Buildings at Melrose* by Clementine Hunter. Collection of Thomas N. Whitehead

SITE LOCATIONS AND CONTACTS

(These sites vary widely in accessibility: some are open daily, some seasonally, some by appointment only.)

Chapter One

Tlingit Clan House
Wrangell, AL 99929
Nora Black-Rinehart (907-874-2023)
Marge Byrd (907-874-3575)

The Gamble House
4 Westmoreland Place
Pasadena, CA 91103
626-793-3334 (www.gamblehouse.org)

Robert Frost Farm
New Hampshire Route 28
Derry, NH 03038 603-432-3091

Chapter Two

Adams National Historic Site
(National Park Service [NPS])
135 Adams Street
Quincy, MA 02169-1749
617-773-1177

James Madison's Montpelier (National Trust for Historic Preservation)
11407 Constitution Highway
Montpelier Station, VA 22957-0067
540-672-2728 (www.montpelier.org)

Lincoln Home National Historic Site (NPS)
413 South 8th Street
Springfield, IL 62701
217-492-4241 (www.nps.gov/liho)

Chapter Three

Rosedown Plantation and Historic Gardens
12501 Louisiana Highway #10
St. Francisville, LA 70775
504-635-3332

The Underground Railroad Houses:
John Parker
300 Front Street
Ripley, OH 45167
937-392-4044/4188

Reverend John Rankin
6152 Rankin Hill Road
Ripley, OH 45167
937-392-1627

Booker T. Washington's "The Oaks"
Tuskegee Institute National Historic Site (NPS)
1212 Old Montgomery Road
Tuskegee Institute, AL 36088
334-727-3200 (www.nps.gov)

Chapter Four

Wukoki
Wupatki National Monument (NPS)
Highway 89
30 miles east of Flagstaff, AZ
520-679-2365 (www.nps.gov)

Vann House State Historic Site
82 Highway 225 North
Chatsworth, GA 30705
706-695-2598 (www.ganet.org/dnr)

Petaluma Adobe State Historic Park
3325 Adobe Road
Petaluma, CA 94952
707-762-4871

Hubbell Trading Post National Historic Site (NPS)
Highway 264
Ganado, AZ 86505-0150
520-755-3475 (http://www.nps.gov/hutr)

Chapter Five

Rebecca Nurse Homestead
149 Pine Street
Danvers, MA 01923
978-774-0554 (Richard Trask, Curator)

Hearst San Simeon
Historical Monument
750 Hearst Castle Road
San Simeon, CA 93452-9741
800-444-4445

Eleanor Roosevelt's Val-Kill Cottage
Roosevelt-Vanderbilt National Historic Site (NPS)
Route 9G
Hyde Park, NY 12538
914-229-9115 (nps.gov.elro)

Chapter Six

Biltmore Estate
Asheville, NC 28801
800-543-2961 (www.biltmore.com)

Lower East Side Tenement Museum
66 Allen Street
New York, NY 10002
212-431-0233 (wnet/org.tenement)

'Iolani Palace
364 South King Street
Honolulu, Hawai'i 96813
808-522-0832

Chapter Seven

Edgar Allan Poe's Cottage
Grand Concourse at Kingsbridge Road
Bronx, NY 10467
718-881-8900
(Bronx County Historical Society)

South Pass City State Historic Site
125 South Pass Main Street
South Pass City, WY 82520
307-332-3684

Melrose Plantation
Route 119
near Natchitoches, LA 71457
318-379-0055

The other houses in the book are privately owned and not open to visitors.

INDEX